Feng-Hsing Wang, Jeng-Shyang Pan, and Lakhmi C. Jain

Innovations in Digital Watermarking Techniques

Studies in Computational Intelligence, Volume 232

Editor-in-Chief

Prof. Janusz Kacprzyk
Systems Research Institute
Polish Academy of Sciences
ul. Newelska 6
01-447 Warsaw
Poland
E-mail: kacprzyk@ibspan.waw.pl

Feng-Hsing Wang, Jeng-Shyang Pan,
and Lakhmi C. Jain

Innovations in Digital Watermarking Techniques

Dr. Feng-Hsing Wang
Department of Electronic Engineering
National Kaohsiung University of Applied
Sciences
415 Chien-Kung Road
Kaohsiung 807
Taiwan
E-mail: jspan@cc.kuas.edu.tw

Prof. Lakhmi C. Jain
University of South Australia
Adelaide
Mawson Lakes Campus
South Australia SA 5095
Australia
E-mail: Lakhmi.jain@unisa.edu.au

Prof. Jeng-Shyang Pan
Department of Electronic Engineering
National Kaohsiung University of Applied
Sciences
415 Chien-Kung Road
Kaohsiung 807
Taiwan
E-mail: jspan@cc.kuas.edu.tw

ISBN 978-3-642-26038-4   e-ISBN 978-3-642-03187-8

DOI 10.1007/978-3-642-03187-8

Studies in Computational Intelligence   ISSN 1860-949X

© 2009 Springer-Verlag Berlin Heidelberg
Softcover reprint of the hardcover 1st edition 2009

Typeset & Cover Design: Scientific Publishing Services Pvt. Ltd., Chennai, India.

Printed in acid-free paper

9 8 7 6 5 4 3 2 1

springer.com

Preface

With the widespread use of the Internet and the rapid development of digital technology, more problems such as information security and copyright protection are encountered in the digital world. Among the solutions for these problems, digital watermarking is one of the popular techniques which has been investigated widely by researchers. Many different kinds of watermarking schemes have been proposed. They provide varied features and functions which can be employed to solve the problems encountered. This research book aims to introduce some novel digital watermarking techniques for digital data stored in image formats, and to introduce intelligent techniques into the existing watermarking systems for performance improvement.

This book is divided into four parts. In the first part, the motivation for this book and the importance of digital watermarking and intelligent technology are described. The organisation of this book is also introduced. In the second part, different kinds of watermarking schemes based on spatial domain, transform domain, and vector quantisation domain are introduced. Also, the group optimization techniques are considered for those mentioned watermarking systems. Some training procedures using genetic algorithms or tabu search are illustrated. By introducing the training procedure into the existing watermarking systems, the performance, such as imperceptibility, capability, or robustness, can be improved. Simulation results are presented to show their superiority. In the third part, several hybrid watermarking systems are illustrated. They demonstrate that by taking the consideration of digital watermarking into account, traditional image coding systems can then possess the ability of digital watermarking. Finally, in the last part, we conclude this book and propose some possible direction for future study.

This book cannot be completed without the help, suggestions, encouragement, and support from many people. We are grateful to Professor C. P. Lim for contributing Sections 3.2 and 3.3. We also wish to thank Dr. H. C. Huang for contributing Section 3.5 and parts of the contents of Chapters 4 and 9. Their contributions do enrich the content of this book.

We wish to thank Springer-Verlag for the professional support.

We wish to thank the reviewers for their visionary feedback as well as the SCI Data Processing Team of Scientific Publishing Services for their excellent work during the preparation of the manuscript. Our sincere thanks to everyone else involved in the completion of this book.

Feng-Hsing Wang
Jeng-Shyang Pan
Lakhmi C. Jain

Contents

Part IV: Summary

List of Abbreviations

ANN = Artificial Neural Network

ART = Adaptive Resonance Theory

BCR = Bit Correct Rate

BER = Bit Error Rate

DCT = Discrete Cosine Transform

DWT = Discrete Wavelet Transform

EA   = Evolutionary Algorithm

EANN = Evolutionary Artificial Neural Network

ED   = Euclidean Distance

GA   = Genetic Algorithm

GCP = Genetic Codebook Partition

GIA = Genetic Index Assignment

GPS = Genetic Pixel Selection

GSVQ = Gain-Shape Vector Quantisation

GWM = Genetic Watermark Modification

JPEG = Joint Photographic Experts Group

LBG = Linde-Buzo-Gray (Algorithm)

LSB = Last Significant Bit

MDC   = Multiple Description Coding

MDSQ = Multiple Description Scale Quantisation

MDVQ = Multiple Description Vector Quantisation

MSE   = Mean Squared Error

MSVQ = Multi-Stage Vector Quantisation

NC   = Normalized Correction

NN   = Neural Network

PSNR = Peak Signal-to-Noise Ratio

QF   = Quality Factor

SNR   = Signal-to-Noise Ratio

SOM   = Self-Organising Map

TS   = Tabu Search

VC   = Visual Cryptography

VQ   = Vector Quantisation

VSS   = Visual Secret Sharing

Part I
Introduction and Background

Chapter 1
Introduction

With the rapid development of digital technology, the treatments for digital data
such as copyright protection and ownership demonstration are becoming more
and more of greater importance. Due to this, the motivation and the goals of
this book are to introduce some watermarking methods for solving the problems
encountered. In this chapter, the reasons why digital watermarking and intelligent techniques are significant in providing solutions for problems in the current
digital world are introduced. The motivation, the aims, and the organisation of
this book are then introduced in the following sections. Some test images used
in the experiments and some notes regarding the introduced systems are also
given in this chapter.

1.1 Importance of Digital Watermarking

Today storing information and data such as documents, images, video, and audio
in digital formats is very common. For many people transferring digital files
via the Internet is a daily activity. Owing to the rapid development of digital
technology and the widespread use of the Internet, life becomes increasingly
more convenient than previously. However, accompanying this convenience, more
serious problems are more prevalent.

As is well known, due to the nature of digital information, it is easy to make
unlimited lossless copies from the original digital source, to modify the content,
and to transfer the copies rapidly over the Internet. Therefore, the demands
of copyright protection, ownership demonstration, and tampering verification
for digital data are becoming more and more urgent. Among the solutions for
these problems, digital watermarking [3] [11] [47] [67] is the most popular one.
Researchers have given consideration to this in the past decade.

Digital watermarking is a process that embeds or inserts extra information,
named the *watermark* or *mark*, into the original data to generate the output,

F.-H. Wang, J.-S. Pan, and L.C. Jain: Innovations in Dig. Watermark. Tech., SCI 232, pp. 3–10.
springerlink.com   © Springer-Verlag Berlin Heidelberg 2009

which is called a *watermarked* or *marked* data. In this process, usually one user-key is required. By the use of the key, the watermarking system can (i) tell the users whether a suspicious source contains the watermark or not, which is known as *detection*, or (ii) reveal the hidden watermark to the users, which is known as *extraction*. In other words, without the correct key used in the encoding process, which is also known as *embedding*, no useful information can be extracted.

There are many kinds of watermarking techniques. Each of them provides different features and functions which can be employed for different purposes. For instance, for tampering verification, a sender of a digital article can put a watermark into the article by employing the *fragile* watermarking method before sending it. When the article is received, an attempt is made to extract the hidden watermark. If no watermark can be extracted, it means the article received has been tampered with. Other examples and applications of digital watermarking are introduced in Chap. 2. They will highlight the importance of digital watermarking.

1.2   Importance of Intelligent Technology

Learning from experience is one very important characteristic of human beings. People avoid making wrong decisions according to their experience, although sometimes the decisions based on experience may turn out to be incorrect. No matter how, people do gain new experience from marking decisions and the accompanying results, even the decisions they made are not good. The accumulation of experience will be used again when making the next decisions for similar encounters.

Intelligent technology [1] [17] [41], which possesses similar qualities as human behavior such as learning from experience, has been developed and applied in many engineering fields (for example, data mining) successfully. Obviously, by introducing intelligent techniques in watermarking systems, it is expected that their performance can be improved.

The field of intelligent technologies contains many interesting subjects such as fuzzy theory and neural networks. The focus of this book is on the use of optimization techniques. The most popular and well-known genetic algorithms (GAs) [21] is applied to develop some genetic watermarking techniques. Also, tabu search [49], which possesses similar structure, will be introduced too. Details of them are illustrated in Chap. 3. Relating training procedures proposed in the literature for watermarking techniques are presented in Part 2.

1.3   Aims and Organisation of This Book

As mentioned previously, digital watermarking techniques can be applied for solving problems such as copyright protection. Therefore the main goals of this book are to introduce some watermarking schemes and to introduce how to improve the performance of them. This book is divided into four parts. The first part includes three chapters. Chapter 1 gives the general introduction about the importance of digital watermarking and intelligent techniques. It also describes the motivation and organisation of this book. Chapter 2 introduces the background knowledge of digital watermarking techniques for still images. Chapter 3 introduces the general ideas of some intelligent techniques, which are neural networks, evolutionary artificial neural networks, genetic algorithms (GAs), and tabu search. With the basic knowledge of digital image watermarking and the intelligent techniques introduced, we are able to proceed further.

In the second part of this book, some digital watermarking schemes based on different domains are introduced. And, some training procedures using the techniques mentioned in Chap. 3 are introduced into these watermarking schemes for better performance. We begin with introducing two spatial-domain-based schemes and a genetic pixel selection procedure in Chap. 4. A transform-domain-based scheme with a genetic band selection procedure is given in Chap. 5. In Chap. 6, some schemes based on the vector quantisation domain are investigated. Then, a training procedure named genetic codebook partition is introduced to improve their performance. In Chap. 7, an index assignment procedure for general watermarking systems is presented. This procedure enhances general watermarking schemes the ability of embedding gray-scaled watermarks. In Chap. 8, two watermark modification procedures, which employ GA to find a better way to modify the original watermarks, are described. By introducing the procedure into general watermarking systems, the embedding results can have better robustness, imperceptibility, or capacity. After the introduction of these watermarking systems and the demonstration of the simulation results presented, readers should have had the concepts about how a training procedure can provide for an existing watermarking system. Further more, readers should be able to know how to design a training procedure.

In the third part of this book, our focus is shifting to introducing watermarking into some existing image coding systems. We believe that by doing so the original coding systems can then provide new features for solving some existing problems such as copyright protection or those mentioned in Sect. 2.3. We illustrate several hybrid watermarking systems in this part. In Chap. 9, we firstly introduce the multiple description coding (MDC) system and then the MDC-based watermarking scheme. With the aid of watermarking, the coded data can now be transmitted over noisy channels with stronger resistance to noise. Also, in this hybrid system, the concetp of applying a training procedure, as introduced in Part 2, is considered. Here the tabu search is employed in the training procedure to find a better way to split the codebook used. In Chap. 10, a multi-stage VQ (MSVQ) based watermarking system is presented. Here an index modification method is employed to handle the coded results of MSVQ, so that the original MSVQ system can provide the ability of watermarking. Further more, in this chapter the concept of embedding fake watermarks is considered. We believe that with the use of these recognizable fake watermarks, the security of an existing watermarking system can be upgraded. In Chap. 11, visual cryptography (VC) and the VQ-index modification method mentioned above are combined together and appended to a gain-shape VQ (GSVQ) system. Here the index modification method plays the role of granting the existing GSVQ system the ability of watermarking, and VC plays the role of securing the hidden watermark. This hybrid system can be implemented without altering the architecture of the original GSVQ system. This means we can just add the extra parts to enhance the original system the extra abilities, such as watermarking.

Finally, the last part summarises this book and points out the essence of this work. It also foreshadows the possible directions for future research. We do hope that this book could help readers to build up the basic concepts of image watermarking and to gain the ideas about how intelligent techniques can play in the area of watermarking. Further more, it would be a great achievement of this book: to raise the interests of readers in the topic of watermarking!

1.4   Test Images Used in Simulation

The watermarking schemes introduced in this book are all image-based techniques. So, to test the performance of them, a number of experiments had been done. For convenience, the images used frequently are listed here.

The binary images displayed in Fig. 1.1 with size 128 × 128 pixels were used as the watermarks.

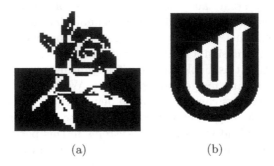

(a)   (b)

Fig. 1.1. The binary watermark images used in experiments. (a) ROSE, and (b) the logo of UniSA.

The gray-scaled image shown in Figs. 1.2 to 1.4 were used as the cover images. We name them LENA, PEPPERS, and BABOON respectively. The size of either image is 512 × 512 pixels.

Fig. 1.2. The image of LENA

Fig. 1.3. The image of PEPPERS

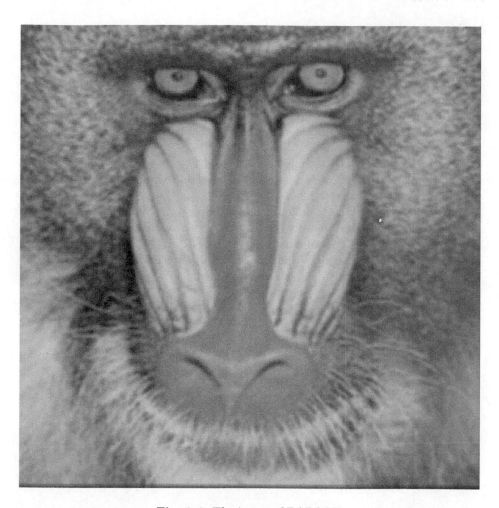

Fig. 1.4. The image of BABOON

1.5   Note

At this point the reader is reminded that:

  (i) The digital watermarking techniques presented in this book are designed for digital images only. That is, either the original source or the watermark is in digital image format. However, the techniques presented can still be modified for different data formats.

 (ii) All the watermarking techniques presented belong to the *invisible* watermarking techniques. More details about invisible watermarking techniques can be found in Sect. 2.2.1 of Chap. 2.

(iii) All the watermarking techniques presented belong to the *readable* watermarking techniques. More details about readable watermarking techniques are presented in Sect. 2.2.2 of Chap. 2.

 (iv) All the watermarking techniques presented are *non-reversible* watermarking techniques. More details relating to non-reversible watermarking are presented in Sect. 2.2.8 of Chap. 2.

  (v) The term "watermarking" denotes "digital watermarking" in the whole book except the extra description mentioned.

Chapter 2
Digital Watermarking Techniques

In this chapter, some basic concepts about digital watermarking are described in as general terms as possible. An introduction is given first. The classification of watermarking, the functions provided by watermarking, and the benchmarks and evaluating measurements for watermarking systems are then introduced in the following sections.

2.1 What Is Digital Watermarking?

Watermarking is an old technique. It had been used widely in the past. A traditional and well-known example is the use of invisible ink. People wrote secret information using invisible ink in order to avoid detection from prying eyes. From a general point of view, the definition of watermarking may be thought of as a method to insert or embed extra information into the media, and also to indicate the method used to obtain the embedded information. The general definitions of some common terms used in the area of watermarking are listed below.

Watermark: The information to be hidden. The term *watermark* also contains a hint that the hidden information is transparent like water.

Cover Media: The media used for carrying the watermark. Sometimes the terms *original media* and *host media* are used to express it.

Watermarked Data: The media which contains the watermark.

Embedding: The procedure used for inserting the watermark into the cover media.

Extraction: The procedure used for extracting the embedded watermark from the watermarked data.

Detection: The procedure used for detecting whether the given media containing a particular watermark.

Watermarking: The method which contains the embedding operator and the extraction/detection operator.

Noise: The natural noise occurred to the watermarked data during transmission.

F.-H. Wang, J.-S. Pan, and L.C. Jain: Innovations in Dig. Watermark. Tech., SCI 232, pp. 11–26.
springerlink.com

Attack: The artificial processes used for modifying the watermarked data in order to destroy the watermark contained in the watermarked data.

Attacked Data: The watermarked data which contains natural noise and/or artificial modification.

In the past decade, owing to the rapid-development of computer technology, people had shifted their focus from traditional media to digital media. As a result, watermarking techniques for digital data have been developed and have become popular.

2.1.1 Elements of a Watermarking System

According to [3], a watermarking system is regarded as a communication system consisting of three main parts: a transmitter, a communication channel, and a receiver, as illustrated in Fig. 2.1.

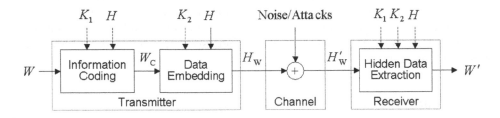

Fig. 2.1. Elements of a watermarking system

The information-coding procedure encodes, compresses, and/or encrypts the original watermark W according to a user key K_1. The data-embedding procedure then embeds the encoded result W_C into the host data H according to another user key K_2. The watermarked data H_W is then delivered to the receiver via some kind of channel. During the transmission, some natural noise or artificial attacks may occur, as a result the received data H'_W of the receiver may be different from the output data H_W of the transmitter. To recover the information hidden in H'_W, the hidden-data-extraction procedure is executed. In the above systems, ether K_1, K_2, or H may or may not have to be presented, according to the algorithms used.

2.2 Types of Digital Watermarking Techniques

In this section, all the existing watermarking techniques are classified into several categories according to different points of view [3] [11] [47] [59] [85].

2.2.1 Visible and Invisible

From the view point as to whether the embedded watermark can be seen by bare human eyes or not, all watermarking techniques can be classified as visible techniques and invisible techniques. For example, Fig. 2.2(c) shows a picture which contains a visible logo of University of South Australia (in short, UniSA) in its top-left corner and Fig. 2.2(d) shows a picture which contains an invisible watermark.

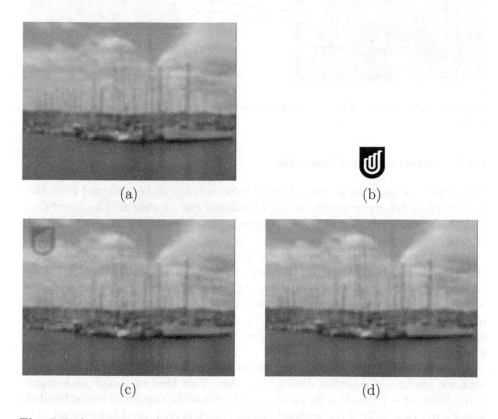

Fig. 2.2. An example of applying the visible watermarking technique and invisible watermarking technique to a given picture. (a) The original picture. (b) The watermark. (c) The watermarked picture containing a visible logo in its top-left corner. (d) The watermarked picture containing an invisible watermark therein.

Obviously, at least two disadvantages exist in visible watermarking techniques:

(i) The visible watermark is not difficult to be removed. The methods proposed in [38] or the image-cropping schemes illustrated in Fig. 2.3 can be used for example.

(ii) The visible watermark degrades the visual quality of the host picture.

In the invisible type of watermarking techniques, the embedded watermark is invisible. It is difficult to distinguish between the original image and the watermarked image. Thus, it is not easy to remove or destroy the embedded watermark without degrading the visual quality of the watermarked image significantly.

Fig. 2.3. An example of removing the embedded watermark from the watermarked picture

2.2.2 Detectable and Readable

From the view point as to what kind of information can be obtained from the watermarked data, the watermarking techniques can be classified as detectable techniques and readable techniques. Figure 2.4 illustrates the general distinction between the two types of techniques.

In the detectable type of techniques, one can only verify if a specified signal (the watermark) is contained in the cover work. In other words, the detectable type of systems only gives a binary answer: yes or no. In contrast, the readable watermarking systems extract and reveal the embedded watermark. As described in Sect. 2.1, researchers use *detection* to illustrate the process of obtaining a binary answer, and *extraction* to express the process of revealing the hidden watermark.

For those techniques which belong to the detectable type, the embedded watermark has to be presented during detection. This kind of technique is more private since it is impossible for an attacker to guess the content of the embedded watermark, especially if the embedded watermark is encrypted beforehand.

2.2.3 Spatial, Transform, and Quantisation

The question of where to hide the signal of the watermark may be divided into three categories: spatial-domain-based techniques, transform-domain-based techniques, and quantisation-domain-based techniques.

Generally speaking, the main concept of spatial-domain-based techniques [4] is to modify the raw data (pixels) of the original host image directly when hiding the watermark bits. The traditional method is to change the Last Significant Bits (LSB) of certain pixels of the host image according to the watermark bits (Sect. 4.2.1). For transform-domain-based techniques, the raw data of the host image are first transformed into frequencies using the discrete cosine transform (DCT) [2], the discrete wavelet transform (DWT) [87], or other types of transforms. These frequencies are then modified according to the watermark bits so that the goal of data hiding can be achieved. Then, the inverse transform is

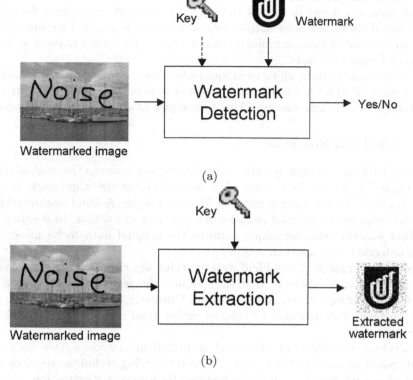

Fig. 2.4. The detectable and the readable watermarking systems

executed and a watermarked image is formed. The quantisation-domain-based techniques, such as vector quantisation (VQ) [24], first quantify the host image using the predefined code-vectors. The indices obtained are then modified according to the watermark bits. The recovery process is finally performed to reconstruct a watermarked image from these modified indices.

Comparing with the three types of techniques, spatial-based techniques possess the advantages including easy implementation, better visual quality, and shorter coding time. However, they also have the disadvantages such as weak robustness for example. The transform-based techniques usually have better robustness and good visual quality in watermarked result. However, they consume more time in the transform and inverse-transform procedures. For the quantisation-based techniques, the most significant feature is they enhance the traditional quantisation systems the watermarking ability.

2.2.4 Robust, Semi-Fragile, and Fragile

Watermarking techniques can also be classified as *robust*, *semi-fragile*, and *fragile* techniques, according to whether the techniques have strong resistance to natural noise and/or to artificial modification (named *attack*). If a watermarking technique can detect or extract the hidden watermark successfully from the watermarked data when noise and/or attack occurred, it is called a robust technique. In contrast, a watermarking technique that cannot resist noise or attacks is called a fragile technique.

There are some watermarking techniques which have strong resistance to some kinds of noise or attack but have weak resistance to other kinds of noise or attack. Researchers named these watermarking techniques as semi-fragile techniques.

2.2.5   Blind and Non-blind

If a watermarking technique resorts to the comparison between the original non-watermarked data and the watermarked one to recover the watermark, it can be classified as *blind* technique and *non-blind* technique. A blind watermarking technique requires no original data for detection or extraction. In contrast, a non-blind watermarking technique requires the original data to be presented during detection or extraction.

In real-world practices, non-blind watermarking algorithms are unsuitable for many practical applications in that they require the non-watermarked data to be presented during extraction or detection. Currently most researchers are focusing on blind watermarking techniques rather than non-blind watermarking techniques.

In addition, definitions of blind and non-blind in nowadays have been extended. Some researchers think that if a watermarking technique requires the present of any information used in embedding for watermark extraction, it then should be classified as non-blind. Based on this definition, one watermarking technique which requires no non-watermarked data but requires the knowledge of embedding position when extraction, is regarded as a non-blind technique.

2.2.6   Public and Private

A watermark is named *private* if only the authorized users can recover it. In other words, it is impossible for unauthorized people to extract the information hidden within the host data. By contrast, a watermarking technique that allows anyone to read the embedded watermark is referred as a *public* watermarking technique.

From the view point of information theory, security cannot be based on algorithms but rather on the choice of the user key. Therefore, researchers believe that private watermarking techniques have superior robustness when compared to public watermarking techniques.

2.2.7   Symmetric and Asymmetric

A watermarking algorithm is called *symmetric* if the detection/extraction process makes use of the same set of parameters used in the embedding process. Here the parameters include the secret keys and other information which may be used to define the embedding position and the embedding process. In contrast, a watermarking algorithm is said *asymmetric* if it uses different keys and parameters for the embedding and the detection/extraction operations.

Researchers believe, for symmetric watermarking techniques a knowledge of these parameters is likely to give pirates enough information to remove the watermark from the watermarked data. Therefore, increasing attention has been given to asymmetric watermarking schemes. Generally speaking, asymmetric watermarking algorithms use a private key for watermark embedding and use a public key for watermark detection/extraction.

2.2.8 Reversible and Non-reversible

A watermarking algorithm is said *reversible* if the watermarked signal can be converted to a non-watermarked signal after the embedded watermark is extracted. By contrast, watermarking algorithms that can not convert the watermarked signal to a non-watermarked signal are named *non-reversible* watermarking algorithms.

Currently, most of the existing watermarking algorithms are non-reversible algorithms, since the selected signals of the cover media have been changed permanently for carrying the watermark bits.

2.3 What Can Digital Watermarking Do?

In the past decade, digital watermarking had been investigated and utilized for a number of different purposes. In this section, we present some common areas served by digital watermarking in order to highlight its usefulness.

2.3.1 Data Hiding

As described previously, digital watermarking can be used for data hiding. For example, a secret message such as "Meet at 13:30, old place" may be regarded as the watermark and hidden in a cover picture, as shown in Fig. 2.5. The watermarked picture is then delivered to the receiver without becoming obvious.

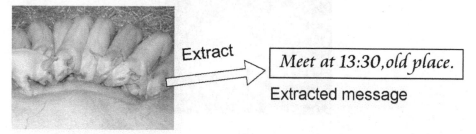

Extract

Meet at 13:30, old place.

Extracted message

Watermarked image

Fig. 2.5. An example of using digital watermarking for data hiding. The secret message "Meet at 13:30, old place" is embedded in the cover image.

2.3.2   Information Integration

People sometimes wish to attach information such as the title, the date, and the place to a digital photo for example. This information is sometimes stored in separate digital files attached to the original photo, and is sometimes saved within the photo directly. Figure 2.6 shows the examples of these two methods used on a digital photo.

For the first method, the extra separate file sometimes causes problems such as not easy to maintain. In the second method, the added comments sometimes cause unacceptable degradation. To solve these problems, invisible and readable watermarking techniques can be employed.

Similar to the first application that uses watermarking for the purpose of data hiding, the comments or notes for the photo are regarded as a watermark. It is then embedded into the original photo using the invisible and readable watermarking technique. This achieves the goal of information integration, as illustrated in the example shown in Fig. 2.7.

Note, different from saving the extra information in the header of the picture file (for example, photo saved in JPEG format allows extra information to be stored in its EXIF section of the header), digital watermarking techniques embed the information within the graphic data itself.

Original photo   Text file

```
File:   A0001.jpg
Date:   20/6/2003
Title:  Adelaide
```

(a)

(b)

Fig. 2.6. Methods used for commenting a digital photo: (a) Saving the information as a separate text file, and (b) adding the information directly to the photo, where the pixels under the text are then obscured.

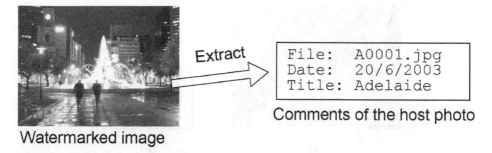

Fig. 2.7. The example of information integration using the invisible and readable watermarking technique. The information for the photo is embedded within the photo.

2.3.3   Intellectual Property Rights (IPR) Protection

Due to the popularity of the Internet, the rapid-development of digital techniques, and the easy-distribution of digital assets, digital watermarking is used to protect intellectual property rights. The creators or legitimate owners of the digital works who want to protect their rights can insert watermarks within their works. Users who plan to make illegal copies of the watermarked works will not be succeed because the machines detect the copyright watermarks and abort the copying procedure.

To implement this concept upon digital products such as DVD or computer software, some related issues such as industry establishment of standards, legislation, and cooperation of hardware manufactures, have to be established. Currently, many issues relating to IPR protection are still proceeding.

2.3.4   Ownership Demonstration

Another classical use is ownership demonstration. The authors of the digital works can embed their own logos or marks within their works to show who the creators or owners are, no matter whether their watermarked works have been modified or not. An example shown in Fig. 2.8 demonstrates such an application. In this example, the creator of Fig. 2.8(a) embeds his name (Fig. 2.8(b)) within this photo. Afterwards, even someone modifies this watermarked photo, the creator can still extract his name from the attacked photo to show the creator of the photo is H. Poirot (Fig. 2.8(c)). In addition, due to the watermarked image has been modified, the extracted watermark therefore contains distortion.

2.3.5   Tampering Verification

Today modifying or tampering with a digital source using computers is not difficult. Consequently, information technology experts warn people not to trust completely what they see in digital form. For the users who want to know whether the data are trust worthy, fragile watermarking techniques provide a possible solution.

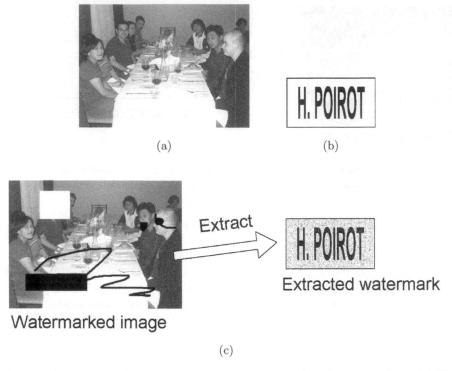

Fig. 2.8. An example of ownership demonstration using digital watermarking. (a) The original image. (b) The watermark which indicates the author or owner. (c) The watermark extracted from the watermarked image.

For example if say Bob wants to send a digital file to Alice. He embeds a fragile watermark in the file and delivers it to Alice by a channel which could be the Internet. Before Alice receives the file, John happens to obtain the watermarked file. He modifies the content of the file and sends it to Alice afterwards. When Alice receives the corrupted file she has no idea as to whether the content is dependable. She therefore verifies if the received file contains the watermark. Due to the fragile nature of the watermark, it has weak resistance against tampering, Alice is unable to find any watermark. She knows immediately that the file she has received had been tampered with.

2.4 Benchmarks and Evaluating Functions

In this section, the benchmarks and evaluating functions for digital watermarking techniques are introduced. The brief introduction for some developed tools and software is also given.

2.4.1 Benchmarks

To evaluate whether a watermarking algorithm has good performance or not, the points outlined in the following sections are usually considered.

A. Imperceptibility

Imperceptibility, or *transparency*, refers to the visual quality of the watermarked result. As stated previously, digital watermarking modifies the host data so that the watermark is not visible to the normal observer. Obviously, there must be some distortion to the host signal caused by the modification. Researchers therefore consider a watermarking algorithm is good if the distortion is minimal.

For image watermarking, the most well-known functions for evaluating visual quality are: the Euclidean distance (ED), the mean-square error (MSE), the normalized correction (NC), and the peak-signal-to-noise ratio (PSNR). The definitions of these functions are given in Eqs. (2.1)–(2.4) respectively.

$$ED(\mathbf{X}, \mathbf{X}') = \sum_i^M \sum_j^N (X(i,j) - X'(i,j))^2. \tag{2.1}$$

$$MSE(\mathbf{X}, \mathbf{X}') = \frac{ED(\mathbf{X}, \mathbf{X}')}{M \times N}. \tag{2.2}$$

$$NC(\mathbf{X}, \mathbf{X}') = \frac{\sum_i^M \sum_j^N (X(i,j) \times X'(i,j))}{\sum_i^M \sum_j^N (X(i,j)^2}. \tag{2.3}$$

$$PSNR(\mathbf{X}, \mathbf{X}') = 10 \times \log_{10} \frac{255^2}{MSE(\mathbf{X}, \mathbf{X}')}. \tag{2.4}$$

Here $\mathbf{X}$ and $\mathbf{X}'$ denote the original image and the processed image, M and N denote the width and height of the images, and $X(i,j)$ and $X'(i,j)$ denote the pixel at position (i,j) of $\mathbf{X}$ and $\mathbf{X}'$ respectively.

In addition, for those digital images in binary format, the Hamming distance (HD), the bit error rate (BER), and the bit correct rate (BCR) can be used:

$$HD(\mathbf{Y}, \mathbf{Y}') = \sum_i^M \sum_j^N |Y(i,j) - Y'(i,j)|. \tag{2.5}$$

$$BER(\mathbf{Y}, \mathbf{Y}') = \frac{HD(\mathbf{Y}, \mathbf{Y}')}{M \times N} \times 100\%. \tag{2.6}$$

$$BCR(\mathbf{Y}, \mathbf{Y}') = (1 - \frac{HD(\mathbf{Y}, \mathbf{Y}')}{M \times N}) \times 100\%. \tag{2.7}$$

In these equations, $\mathbf{Y}$ and $\mathbf{Y}'$ denote the original image and the processed image, M and N denote the width and height of the images, and $Y(i,j)$ and $Y'(i,j)$ denote the pixel at position (i,j) of $\mathbf{Y}$ and $\mathbf{Y}'$ respectively.

B. Robustness

In some applications, the ability of the embedded watermark to survive noise or attacks is important. To determine whether a watermarking algorithm has adequate robustness, researchers usually employ a number of processes such as image processing methods to attack the watermarked data and to observe if the embedded watermark can be detected or extracted successfully.

From the experimental results provided in literature research, a greater robustness usually means the visual quality of the watermarked result is degraded.

C. Capacity

Capacity denotes the number of bits the watermarking algorithm can embed within the host data. Obviously, the more bits embedded, the better capacity the watermarking algorithm has. However, it is believed that embedding more bits results in poorer visual quality of watermarked images.

D. Security

Security is used for expressing how well the hidden watermark can be protected. Usually, researchers evaluate the security of a watermarking algorithm by calculating using how many CPU cycles or how long it takes to break the watermarking algorithm or to reveal the hidden watermark.

E. Coding Time

Coding time denotes the time consumed by the embedding procedure and the detection/extraction procedure. In some cases, people are more concerned over the detection/extraction time than the embedding time, since the embedding procedure may only be performed once.

F. Summary

There are other points which can be used for evaluating the performance of a watermarking algorithm. These include the material such as that contained in References [3] and [62]. Generally, researchers will not determine a watermarking algorithm to be good or bad only from a consideration of the points listed above. Instead, they prefer to consider other factor such as what application the watermarking algorithm is served. For example, for the purpose of tampering verification (Sect. 2.3.5) using digital watermarking, a robust watermarking scheme becomes useless since tampering verification requires the feature of fragile.

2.4.2 Attack Schemes

Robustness is one of the important features of digital watermarking. Thus, we introduce the common methods used for attacking watermarked images.

A. Compression

Image compression is a common method used for attacking watermarked images. Usually image compression algorithms remove the redundant signals (e.g., high frequency) from the input images to achieve data compression. Consequently, it is useful when attacking watermarked images.

In addition, for image processing, the JPEG compression [6] [12] and the VQ compression [24] are the most well-known and commonly used.

B. Spatial Filtering

Using spatial masks for image processing is usually called *spatial filtering*. The masks themselves are called *spatial filters* [27]. The basic approach when using spatial filtering is to sum products between the mask coefficients and the luminance of the pixels under the mask at a specific location in the image.

The most common filters are low-pass filtering, high-pass filtering, and median filtering. The masks of window size $= 3$ for low-pass filtering and high-pass filtering are expressed by Eqs. (2.8) and (2.9) respectively.

$$\frac{1}{9}\begin{bmatrix} 1 & 1 & 1 \\ 1 & 1 & 1 \\ 1 & 1 & 1 \end{bmatrix}. \tag{2.8}$$

$$\frac{1}{9}\begin{bmatrix} -1 & -1 & -1 \\ -1 & 8 & -1 \\ -1 & -1 & -1 \end{bmatrix}. \tag{2.9}$$

For median filtering, the luminance of each pixel is replaced by the median of the gray levels in the range of the $(n \times n)$ spatial mask centering around that pixel.

C. Cropping

Another popular scheme of attack is to alter the watermarked images by cropping. In our simulation, for simplicity we crop the bottom-left hand quarter of the watermarked image and replace it with an all-black image.

Researchers in general believe that by applying the concepts of spread spectrum from communications [10] and by making use of the linear feedback shift registers to disperse the spatial domain relationships in the original images [61], the affect of image-cropping schemes may be reduced.

D. Shifting

The attackers may move the watermarked image horizontally and vertically to destroy the watermark information conveyed. If the attacker shifts the water-marked image h pixels to the right and v pixels downwards, the shifted image may be represented by:

$$\mathbf{X}''(i,j) = \mathbf{X}'(i - v, j - h), \tag{2.10}$$

where $\mathbf{X}'$ is the watermarked image and $\mathbf{X}''$ is the shifted result.

Generally speaking, for block-based watermarking algorithms such as the VQ-based or the DCT-based watermarking algorithms, image shifting usually causes the watermark extracting algorithm to lose the synchronization with the water-marked image.

E. Rotation

Rotating the watermarked images is also a common method considered by at-tackers. For a watermarked image $\mathbf{X}'$ with size $(M \times N)$ pixels, the rotated result $\mathbf{X}''$ can be expressed by:

$$\mathbf{X}'' = \bigcup_{i=0}^{M-1} \bigcup_{j=0}^{N-1} \{X''(i,j)\}$$

$$= \bigcup_{i=0}^{M-1} \bigcup_{j=0}^{N-1} \{X'(i\cos\theta, j\sin\theta)\}, \tag{2.11}$$

where θ denotes the rotation angle.

In practical implementations, some parts of the attacked image may not have any value for representing its luminance after rotation. For simplicity, people usually interpolate the neighboring pixels to present the luminance if the miss-ing pixels are inside the watermarked image; otherwise, the luminance of these regions may be set to zero.

F. Summary

Attacking watermarked images or removing the embedded watermark signals from the watermarked images is another topic for robust watermarking algo-rithms. Some researchers have paid attention to this topic and have proposed the attack schemes [54] [89]. Here we have only briefly introduced the common and well-known methods.

It is thought that there is no watermarking algorithm which can thwart all of the attacking schemes. Researchers therefore are more concerned in design-ing and discovering new watermarking algorithms to resist particular kinds of attacks.

2.4.3   Packages and Tools for Evaluation

In this section, some software and computer programs developed to attack watermarked images are introduced. The main aim of the software is to remove the watermark signal from the assigned watermarked images by employing the possible image processing functions, such as those mentioned in Sect. 2.4.2.

There is no known watermarking algorithm which can survive all the attacking schemes, as noted in the summary section of Sect. 2.4.2.

A. Stirmark

The most well-known tool used for attacking watermarked images is Stirmark [63]. It offers many attack schemes including those mentioned in Sect. 2.4.2 and others such as image resizing, luminance scaling, shearing, line deletion for example.

B. Optimark

Optimark [80] is another powerful benchmark package which provides a graphical user interface and similar attack schemes to Stirmark. To use this package, the users first have to assign the embedding and the detection/extraction executable programs, the test images, and other parameters. The attacks will be executed then. Optimark is believed to be the most complete benchmark package so far [67].

C. Checkmark

Checkmark [70] is similar to Stirmark. It incorporates a number of new attacks such as wavelet compression, projective transforms, down/up sampling for example.

D. Certimark

Certimark [64] is popular owing to its open architecture. It allows new functionalities to be integrated easily and provides flexible interface to be plugged-in watermarking software.

E. Image Processing Software

Another method to test robustness is by employing some image processing software such as PhotoShop[1], PhotoImpact[2], or PhotoCap [3] for example. The users must employ the considered image processing schemes provided by the software manually to attack their watermarked images. They then perform their

---

[1] The trademark of this software is belonging to Adobe Inc.
[2] The trademark of this software is belonging to Ulead Inc.
[3] This is a free and green software designed by my colleague Johnson Wang.

watermarking programs to extract/detect the embedded watermarks from the attacked images. By calculating the quality of the extracted/detected results, the users know if their watermarking algorithms have strong robustness against the attacking schemes considered.

2.5 Summary

In this chapter, the background information of digital watermarking such as benchmarks, applications, and classification have been introduced. According to the classifications and applications given in Sects. 2.2 and 2.3, new watermarking schemes can be developed to utilize the properties of the existing watermarking techniques. In addition, using the benchmarks and attack schemes presented in Sect. 2.4, the existing watermarking schemes may be improved and modified for better performance.

Chapter 3
Intelligent Techniques

In this chapter, the reasons for considering optimization techniques in the watermarking research area is explained. The fundamental knowledge of genetic algorithms and tabu search are then introduced. With the knowledge of optimization techniques, we can design some training procedures for watermarking techniques. In Part 2 of this book, some training procedures proposed in the literature for optimizing the developed watermarking schemes will be introduced.

3.1 Introduction

As mentioned in Chap. 1, the use of intelligent techniques provides the ability to solve and improve some of the problems encountered. A variety of intelligent techniques have been proposed for this purpose, which all have been applied successfully in many areas. As a result of the research in the area of watermarking, they have been applied to the design of watermarking systems or in procedures to improve performance.

In [32], Hwang et al. designed a watermarking system using a neural network to provide better imperceptibility and robustness. In [36] [52], the tabu search approach [49] was investigated and used in the watermarking system. In [92], the authors proposed using genetic algorithms (GAs) [21] to select a better pixel set for carrying watermark bits in the spatial domain. In [93], the authors proposed a VQ-based watermarking scheme with a genetic-codebook-partition scheme. They employed GAs to find a better way in which to split the codebook. In [77], Shieh et al. proposed a DCT-based watermarking scheme. They also used GAs to select better DCT bands for watermark embedding. Each watermarking system mentioned above provides better performance after using intelligent techniques in the system.

In order to employ intelligent techniques in the watermarking system and design the training schemes, the fundamental knowledge of some intelligent techniques must be studied first. Among the intelligent techniques, GAs are most popular and well-known. Also, due to the tabu search approach possesses similar framework, we concentrate on them and introduce the fundamental knowledge

F.-H. Wang, J.-S. Pan, and L.C. Jain: Innovations in Dig. Watermark. Tech., SCI 232, pp. 27–44.
springerlink.com

relating to them in this chapter. The following sections present an introduction to the intelligent techniques.

3.2 Neural Network (NN)

In this section, a number of intelligent techniques stemmed from artificial neural network (ANN) models are introduced. First, the self-organising map (SOM) and adaptive resonance theory (ART) networks are presented. Both SOM and ART models and their variants have been successfully applied to a number of watermarking problems [14] [44] [58] [83] [90]. In addition, the procedure for synergizing ANN with the genetic algorithm to form an evolutionary artificial neural network (EANN) is discussed.

3.2.1 The Self-Organising Map (SOM)

The SOM [45] [46] is a neural network model that is used in a wide variety of applications. It is particularly useful for visualisation and clustering of data. By adopting an ordered nonlinear projection technique, the SOM is able to project high-dimensional a set of input data onto a low-dimensional grid.

The SOM consists of N unit grid points (also called nodes) located on a regular low-dimensional grid, usually a two-dimensional grid (or map). Each unit j has an associated d-dimensional prototype vector,

$$\mathbf{w}_j = \{w_{j1}, w_{j2}, ..., w_{jd}\}, \tag{3.1}$$

where j is the index of the node. The lattice type of the array can be defined to be rectangular or hexagonal.

The training procedure of the SOM is as follows. An input vector, $\mathbf{x} \in V$, with $V \subseteq R^d$ at each training step t, is randomly selection from a set of training data. The distances between $\mathbf{x}$ and all prototype vectors are computed. The smallest of the Euclidean distance $\| \mathbf{x} - \mathbf{w}_j \|$ defines the best matched unit (BMU), signified by i^*, i.e.,

$$i^* = \arg\min_i \| \mathbf{x} - \mathbf{w}_i \| . \tag{3.2}$$

The BMU or the winning unit and its neighbours adapt to represent the input even better by modifying their reference vectors towards the current input. The amount the units learn is governed by the unsupervised competitive learning rule:

$$\mathbf{w}_j(t+1) = \mathbf{w}_j(t) + \eta h_{i^*j}(t)[\mathbf{x}(t) - \mathbf{w}_j(t)], \tag{3.3}$$

where the neighbourhood function h is a decreasing function which is related to the distance of the units from the winning node on the grid. If the locations of units i and j on the map grid are denoted by the two-dimensional vectors $\mathbf{r}_i$ and $\mathbf{r}_j$, respectively, then the Gaussian neighbourhood function h is defined as:

$$h_{i^*j}(t) = \exp\left(-\frac{(\mathbf{r}_{i^*} - \mathbf{r}_j)^2}{2(\sigma_\wedge(t))^2}\right), \tag{3.4}$$

with $\mathbf{r}_{i^*}$ and $\mathbf{r}_j$ the lattice coordinates of neuron i^* (winner) and j, of which the range $\sigma_\wedge(t)$ decreases as follows:

$$\sigma_\wedge(t) = \sigma_{\wedge_0}(t) \exp\left(-2\sigma_{\wedge_0}\frac{t}{t_{\max}}\right), \tag{3.5}$$

with t the present time step, $t_{\max}$ the maximum number of time steps, and $\sigma_{\wedge_0}$ the range spanned by the neighbourhood function at $t = 0$. In the original SOM model, the radius of the neighbourhood range $\sigma_{\wedge_0}$ at the beginning of the process may be selected as a fairly large value (rough training), and put to shrink monotonically in further iterations (fine training).

Weight updates in the SOM can be performed using an incremental or "online" learning strategy. However, the training procedure can further be optimized by using batch learning. In batch learning, the weight vectors are updated after all input samples of the training set are considered. In the following section, the batch-learning version of the SOM model is explained.

3.2.2   The Batch-Learning SOM Model

The batch-learning SOM [48] is a variant of the SOM that is based on a fixed-point iteration process. Instead of using one data vector at a time, the whole data set is presented to the SOM before any adjustment of weights takes place. It provides a considerable speed-up to the original SOM training procedure by replacing the incremental weight updates with an iterative scheme that sets the weight vector of each neuron to a weighted mean of the training data. The initial values of the prototype vectors $\mathbf{w}_j$ may be randomly selected, preferably from the input samples [48].

The batch-learning SOM algorithm is as follows. Given a fixed training set $M = \{\mathbf{x}^\mu\}$ of M input samples, $\mu = 1, 2, ..., M$, compare each $\mathbf{x}$ with all $\mathbf{w}_j$, $j = 1, 2, ..., N$, and copy each $\mathbf{x}$ to the sublist associated with the map unit that has the shortest Euclidean distance. When all $\mathbf{x}^\mu$ are distributed into the respective sublist of the map unit (see Fig. 3.1), the new weight vectors are computed according to:

$$\mathbf{w}_j(t+1) = \frac{\displaystyle\sum_{\mu\in M} h_{i^*j}(t)\mathbf{x}^\mu}{\displaystyle\sum_{\mu\in M} h_{i^*j}(t)}, \tag{3.6}$$

where i^* is the BMU of all data vectors as in Eq. (3.2).

The new weight vector is a weighted average of the data samples, where the weight of each data sample is the neighbourhood function value $h_{i^*j}(t)$ at its BMU i^*, i.e.,

$$h_{i^*j}(t) = \exp\left(-\frac{(\mathbf{r}_{i^*} - \mathbf{r}_j)^2}{2(\sigma(t))^2}\right), \tag{3.7}$$

where $\mathbf{r}_{i^*}$ and $\mathbf{r}_j$ represent the lattice coordinates of neurons i^* and j, and $\sigma(t)$ is the neighbourhood range at the t-th training epoch. For every iteration, each

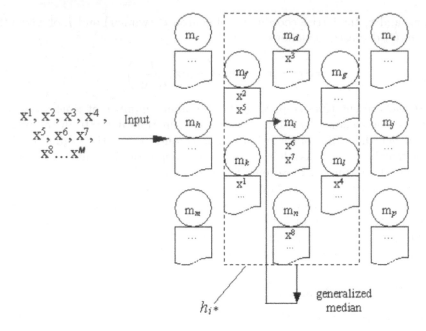

$$x^1, x^2, x^3, x^4, \\ x^5, x^6, x^7, \\ x^8 \ldots x^M \quad \text{Input}$$

Fig. 3.1. Illustration of the SOM with batch learning (adapted from [48])

weight vector is replaced by the generalised median of the input data assigned to the sublist in the map. The procedure is repeated until the maximum number of steps, $t_{\max}$, is reached.

3.2.3   Adaptive Resonance Theory (ART)

ART originated from Grossbergs work [19] [20] on the analysis of cognitive learning processes in humans in 1970s. It attempts to address issues related to human cognitive process, and to devise computational methods mimicking the learning activities of the brain. Since the first inception of the ART-based network known as ART1 [7], a number of ART architectures, both unsupervised and supervised models, have been developed by Carpenter and associates. Similarly to many other ANNs, the growing methodology of ART is based on a similarity measure between the input pattern and learned exemplars, and the one with the highest degree of similarity is selected as the best-matched prototype. Then, an arbitrary threshold is applied to determine whether or not to recruit a new node. Nevertheless, a distinct difference is that ART undergoes a two-stage selection and test process [7] [8]. Figure 3.2 shows a generic architecture of an unsupervised ART model.

On the presentation of a new input pattern, a feedforward pass is exercised to select the most similar prototype using a competitive hypothesis selection process. A feedback pass is then triggered whereby the winning node is used for hypothesis test against a vigilance threshold. If the vigilance test is not

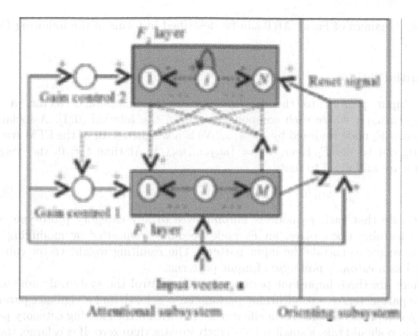

Fig. 3.2. A generic architecture of an unsupervised ART network. Familiar events (patterns) are processed within an attentional subsystem, and an orienting subsystem resets the attentional subsystem when an unfamiliar event occurs. A positive sign (+) indicates an excitatory connection, whereas a negative sign (-) indicates an inhibitory connection (adapted from [7]).

satisfied, then a new round of hypothesis search (selection and test) ensues. The search process continues until the vigilance criterion is satisfied by an existing prototype, or the creation of a new node to include the input pattern. This feedback mechanism, which distinguishes ART from other ANN models, allows ART to form a stable yet adaptive knowledge structure.

3.2.3.1 Fuzzy ART

Fuzzy ART [8] is a generalisation of ART1 for fast unsupervised learning and categorisation of binary and analogue patterns. It achieves a synthesis of fuzzy set theory and adaptive resonance theory by exploiting a close similarity between the computations of fuzzy subsethood and the learning rules of ART1 [7]. In a Fuzzy ART network, the crisp intersection operator ($\cap$) that governs the dynamics of ART1 is replaced by the fuzzy AND or MIN operator ($\wedge$). The MIN operator reduces to the intersection operator in binary cases, thus enabling Fuzzy ART to accept binary as well as analogue input patterns. In comparison with ART1, the differential equations which constitute the learning rules of both the bottom-up and top-down adaptive filters in ART1 reduce to a same form when fuzzy operation is employed. Therefore, Fuzzy ART has only one set of bi-directional weight vectors which subsumes both the bottom-up and top-down weight vectors of ART1.

The dynamics of Fuzzy ART can be described in terms of the following four phases:

A. Initialisation

Each input pattern to the F_1 layer is an M-dimensional vector, $\mathbf{a} = \{a_1, a_2, ..., a_M\}$, where each component a_i is in the interval $[0, 1]$. Associated with each F_2 node (indexed by $j = 1, ..., N$) is a weight vector of the LTM traces fanning out to the F_1 layer, $\mathbf{w}_j = \{w_{j1}, ..., w_{jM}\}$. At time $t = 0$, the weight vectors are initialised to unity

$$w_{j1}(0) = ... = w_{jM}(0) = 1, \, j = 1, ..., N \,, \tag{3.8}$$

to indicate that each F_2 node is *uncommitted* to any category prototype yet. When learning takes place, an F_2 node becomes *committed* by modifying its weight vector to encode the input pattern. The resulting weight vector can be viewed as a *category prototype* of input patterns.

There are three important parameters that control the system dynamics: a choice parameter, $\alpha > 0$; a learning parameter, $\beta \in [0, 1]$; and a vigilance parameter, $\rho \in [0, 1]$ [8]. To ensure an efficient search within the existing category prototypes, α should take a small value slightly greater than zero. If α is large, there is a tendency to choose an new, uncommitted node before going into a deeper search amongst previously committed F_2 nodes [7]. Hence, $\alpha \rightarrow 0$ is known as the conservative limit which allows the system to minimise recoding of similar prototypes. The learning parameter decides the rate at which recoding should occur; whereas the vigilance parameter determines the degree of granularity of the learned categories. Low vigilance values lead to coarse categories with broad generalisation. High vigilance values lead to fine categories with narrow generalisation. By setting $\rho = 0$, the network operates in a "forced choice" condition where it has to make a prediction for every input pattern. On the other hand, by setting $\rho = 1$, prototype learning reduces to exemplar learning where every input pattern is treated as a different category, and is absorbed into the network.

B. Category Choice, Test, and Search

The input vector $\mathbf{a}$ is transmitted to the F_2 layer. The response of each F_2 node is measured using a choice function

$$T_j(\mathbf{a}) = \frac{|\mathbf{a} \wedge \mathbf{w}_j|}{\alpha + |\mathbf{w}_j|} . \tag{3.9}$$

The fuzzy AND operator ($\wedge$) is defined by

$$(\mathbf{u} \wedge \mathbf{v})_i \equiv \min(u_i, v_i) \,, \tag{3.10}$$

and the size, $|\cdot|$, is defined by

$$|\mathbf{u}| \equiv \sum_i |u_i| \,, \tag{3.11}$$

i.e., the L_1 norm. By setting $\alpha \approx 0$, the choice function defined by Eq. (3.9) reduces to a measure of the degree of $\mathbf{a}$ being a fuzzy subset of $\mathbf{w}_j$. For notational simplicity, $T_j(\mathbf{a})$ is written as T_j since $\mathbf{a}$ is fixed. The node that has the highest response, denoted as node J, is selected as the winning node,

$$T_J = \max\{T_j : j = 1, ..., N\}. \tag{3.12}$$

If there is a tie on T_j, then the category with the smallest index is chosen [8]. All other nodes $j \neq J$ are deactivated in accordance with the winner-take-all competition.

The winning node J propagates its weight vector back to F_1. A vigilance test is performed to measure the similarity against the vigilance threshold between the transformed category prototype and the input vector:

$$\frac{|\mathbf{a} \wedge \mathbf{w}_J|}{|\mathbf{a}|} \geq \rho. \tag{3.13}$$

If the vigilance test is satisfied, resonance is said to occur and learning takes place, as defined below. However, if the test fails, node J is inhibited, i.e., it is prohibited from participating in subsequent competitions. Input $\mathbf{a}$ is re-transmitted to F_2 to search for a new winner. This process is repeated, consecutively disabling nodes in F_2, until either a category prototype is able to pass the vigilance test, or, if no such node exists, a new node is created to code the input vector.

C. Learning

Once search ends, learning takes place by adjusting the weight vector, $\mathbf{w}_J$, according to the following learning rule [8]:

$$\mathbf{w}_J^{(\text{new})} = \beta(\mathbf{a} \wedge \mathbf{w}_J^{(\text{old})}) + (1 - \beta)\mathbf{w}_J^{(\text{old})}. \tag{3.14}$$

There are two learning modes: (1) *fast learning* corresponds to setting $\beta = 1$ for all time; (2) *fast-commit and slow-recode learning* corresponds to setting $\beta = 1$ for an uncommitted node, and $\beta < 1$ for a committed node. The former rule allows the weight vector to converge to the asymptotic category boundary in one attempt; whereas the latter rule slowly varies the weight vector to make the system more resistant to noise [8].

D. Complement Coding

It was pointed out that points out that ART1 is sensitive to a category drift and proliferation problem [55]. This problem can be overcome if the norm or size of the input vectors is kept constant [8], i.e.,

$$|\mathbf{a}| \equiv \gamma > 0. \tag{3.15}$$

They also propose a normalisation technique called *complement coding*. With complement coding, an M-dimensional input vector, $\mathbf{a}$, is normalised to a $2M$-dimensional vector, $\mathbf{A}$, as follows

$$\mathbf{A} = (\mathbf{a}, \mathbf{a}^c) \equiv \{a_i, ..., a_M, 1 - a_1, ..., 1 - a_M\}. \tag{3.16}$$

By Eq. (3.16),

$$|\mathbf{A}| = |(\mathbf{a}, \mathbf{a}^c)| = \sum_{i=1}^{M} a_i + (M - \sum_{i=1}^{M} a_i) = M. \tag{3.17}$$

Thus, the size of $\mathbf{A}$ is always made equal to M. When complement-coding is used, the number of nodes in F_1 is doubled, and the weight vector, $\mathbf{w}_j$ has to be extended to

$$w_{j1}(0) = ... = w_{j,2M}(0) = 1, j = 1, ..., N. \tag{3.18}$$

One property of complement-coded inputs is that the j-th prototype node has a hyper-rectangle category boundary, R_j, in a M-dimensional space, and the two vertices of R_j are defined by weight vector, $\mathbf{w}_j$ [8].

3.3   Evolutionary Artificial Neural Networks (EANN)

One of the research domains in ANN is devoted to applying evolutionary algorithms (EAs) to evolve different aspects of ANNs. The combination between ANNs and EAs results in a specific class of ANN known as evolutionary ANN (EANN). EANNs capitalize evolution as another fundamental form of adaptation in addition to learning [106] [107]. The learning adaptation capabilities allow EANN models to perform in a dynamic environment in an effective and efficient manner.

EAs participate in determining the dynamics of ANNs mainly at two levels, i.e., connection weights and architectures. The former normally involves feed-forward ANN models in which their training is based on gradient information such as back-propagation and conjugate gradient. Such gradient-based ANNs are likely to be trapped in a local minimum of the error function [106] [107]. The advantage of EAs is that they do not rely on gradient information to find global solutions. For this reason, EAs have been used to replace the gradient descent algorithm in ANNs for finding optimum weights from multiple regions in the error space [16] [42] [86].

The GA perhaps is the most widely used EA techniques. How, then, the GA is able to enhance the learning capability of an ANN? One of the ways is by searching and adapting the ANN weights using the GA. The idea of such an EANN is illustrated in Fig. 3.3, where the ANN undergoes learning in the network environment and evolution search and adaptation in the GA environment using the training set. The evolved networks are then evaluated using an unseen test data set.

The EANN undergoes adaptation in the *network environment* and the *GA environment*. In the network environment, the ANN performs learning (such as

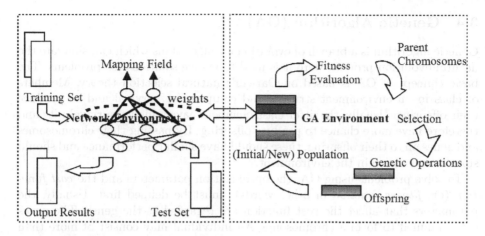

Fig. 3.3. A schematic of hybridisation between ANN and the GA

backpropagation) using the training data set whereas in the GA environment, adaptation is continued by performing a number generation of global search in the data space. The general learning procedure of an EANN is summarised as follows.

(1) **Chromosome representation and population initialisation.** An encoding scheme (either direct or indirect) [106] [107] can be used to encode the weights of the ANN as a chromosome. A finite number of chromosomes are created and these chromosomes are initialised randomly to form a population.

(2) **Fitness evaluation.** Each chromosome in the population is transferred back to the ANN. Each ANN might be trained with an ANN learning algorithm (such as backpropagation). The fitness of the trained ANN is assessed in terms of its performance metrics.

(3) **Chromosomes selection.** A selection scheme (e.g., the roulette-wheel selection, see Sect. 3.4.4) is applied to choose chromosomes based on their fitness values.

(4) **Crossover and mutation.** The genetic operators (i.e., crossover and mutation) are applied to process the selected chromosomes.

(5) **Formation of a new population.** After the crossover and mutation operations, a new set of chromosomes (offspring) is obtained. They are a new set of individuals that replace their parents in a new population.

(6) **Terminating criterion.** The process goes back to Step 3 if a terminating criterion is not satisfied. In other words, the process of fitness evaluation, selection, and genetic operations on a population of chromosomes is repeated until a terminating criterion is satisfied, e.g. a pre-defined number of generations is reached. Once the terminating criterion is fulfilled, the chromosome of the current population that has the highest fitness value is identified as the final weights of the ANN.

3.4   Genetic Algorithm (GA)

Genetic algorithm is a branch of evolutionary algorithms which can simulate the natural evolution process. Here it is used to evolve solutions for problems. The basic concept of GA is based on Darwin's natural selection theory. Members of class in an environment struggle and compete for survival and to produce their offspring. Usually, the fitter class members have a higher survival rate and therefore have more chance to produce offspring. By passing their chromosomes and genes on to their offspring, these should have similar performance and should survive and thrive in the environment.

To solve problems using GA, the optimization parameters and the *cost function* (or, *fitness function* in other words) must be defined first. Usually, the parameters that affect the cost function are regarded as the genes. The genes are combined to form a chromosome. An individual may consist of more than one chromosome. Section 3.4.1 illustrates the relationships among individuals, chromosomes, and genes.

The cost function will usually vary for different cases. For example, in our study of watermarking, imperceptibility usually is an important factor in most schemes. Thus, to optimize imperceptibility for a watermarking scheme, a cost function can be defined as:

$$f = \text{PSNR}(\mathbf{X}, \mathbf{X}').$$
(3.19)

Here PSNR (Eq. (2.4)) is a function used for evaluating imperceptibility, $\mathbf{X}$ is the original image, and $\mathbf{X}'$ is the watermarked image.

Figure 3.4 illustrates the general steps for the GA searching procedure. Details of them are described in the following subsections.

3.4.1   Individual, Chromosome, and Gene

As stated previously, the parameters that affect the cost function are regarded as the genes. We define the chromosome and individual as:

$$\textbf{Individual}_i = \bigcup_{j=1}^{M} \{\textbf{Chromosome}_j\}_i,$$
(3.20)

$$\textbf{Chromosome}_i = \{g_1^i, g_2^i, ..., g_N^i\}, \ i = 1, 2, ..., M,$$
(3.21)

where M is the number of chromosomes contained in an individual, N is the length of a chromosome (or the number of genes contained in a chromosome), and $\{g_1^i, g_2^i, ..., g_N^i\}$ are the genes of the i-th chromosome. Figure 3.5 illustrates their relationship.

3.4.2   Initialization

In this step, the fitness function and some parameters, such as the number of individuals and chromosomes, the selection rate, the crossover rate, the mutation rate, and the considered iteration, for the GA procedure are defined. Also,

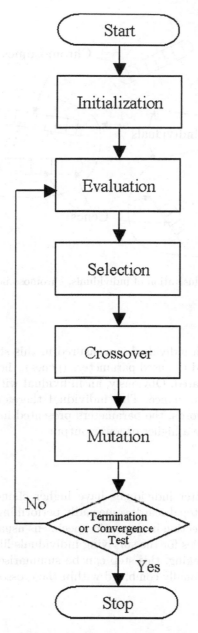

Fig. 3.4. The block diagrams of the GA training procedure

the initial genes are randomly generated within the considered range. The GA is robust [28], and the initial values selected will not affect the trained result significantly.

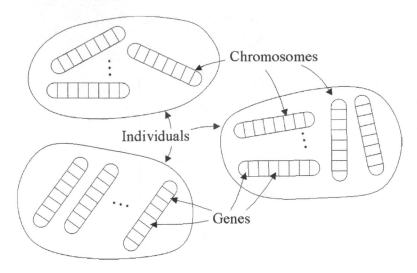

Fig. 3.5. The illustration of individuals, chromosomes, and genes

3.4.3   Evaluation

The performance of each individual is measured in this step. According to the given cost function f and the used parameters (genes), the performance of each individual can be calculated. Obviously, an individual with good genes usually means it has good performance. This individual therefore is more fit in the environment. In other words, the parameters presented in the individual cause the cost function to have a higher or lower output.

3.4.4   Selection

In the natural world fitter individuals have higher chance to survive and to produce offspring. This step discards some badly performing individuals, which is similar to the natural selection in that unfit specimens usually discarded in time. It also assigns probabilities for the surviving individuals likelihood of producing offspring. Generally speaking, this step can be summarised in three sub-steps. Some of them are occasionally combined within the crossover step.

A. Survival Selection

After measuring how fit the individuals are, some of the poorly-performing individuals are discarded in this step. Here let S be the number of individuals and ρ_s be a selection rate which decides how many individuals in current generation survive. According to this, $\rho_s S$ good-performance individuals are selected and the remaining $(1 - \rho_s)S$ individuals are discarded.

Usually, the value of ρ_s varies in different cases. People tend to determine the value according to experience and knowledge.

B. Probability Assignment

The surviving individuals are determined according by their performance $\{f_1, f_2, ..., f_n\}$. Here $n = \rho_s S$ and the probabilities are $\{\rho_1, \rho_2, ..., \rho_n\}$. These are assigned to indicate the probability of producing offspring. Here

$$f = \sum_{i=1}^{n} f_i, \qquad (3.22)$$

$$\rho_i = \frac{f_i}{f} \times 100\%, \; i = 1, 2, ..., n, \qquad (3.23)$$

$$\sum_{i=1}^{n} \rho_i = 1. \qquad (3.24)$$

An example is given in Fig. 3.6, where the third individual has the best fitness score ($f_3 = 27$). The probability calculated by using Eqs. (3.22) to (3.24) is the highest ($\rho_3 = 37.5\%$).

Ind. No.	Fitness score
1	6
2	11
3	27
4	12
5	16

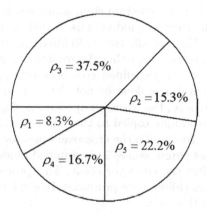

(a) The performance of five individuals   (b) The probabilities assigned

Fig. 3.6. The illustration of probability assignment

C. Parent Selection

After assigning the probabilities of generating offspring to the surviving individuals, a parent-selection procedure is performed to determine which individuals may be selected as *parents*. Here the well-known roulette-wheel-selection scheme is described.

This scheme first generates a random probability. The individual whose segment spans the random number is then selected. An example of this selection scheme is given in Fig. 3.7. In this example, $\rho = 34\%$ is the random-generated probability. Because it is within the segment of the fifth individual, this individual therefore is selected.

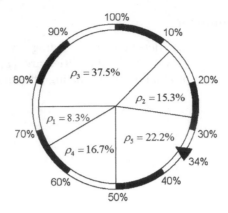

Fig. 3.7. The illustration of roulette-wheel-selection scheme

3.4.5 Crossover

After the selection step, some new individuals, called *children*, are produced by the surviving individuals in this step.

First of all, two individuals from the surviving individuals are selected as parents using either the roulette-wheel-selection scheme or other type of scheme. Then, a predefined crossover rate ρ_c is used to decide whether the crossover operator is done or not. That is, if a random-generated probability is higher than ρ_c, the crossover operator is skipped and the selected parent characteristics are simply copied to the children without change.

To perform the crossover operator, a crossover point between the first and the last chromosomes is determined randomly. Part of each parent is then exchanged after the crossover point. This procedure is illustrated in Fig. 3.8. Afterwards, two children are produced. In some cases, two or more crossover points are used in this step.

The above procedure is repeated until the needed number of child individuals are produced. That is, when there is no individual discarded in the selection step (i.e., $\rho_s = 1$), S child individuals are produced; otherwise, $(1 - \rho_s)S$ child individuals are produced to replace the discarded individuals.

3.4.6 Mutation

In addition to the crossover operator that recombines the chromosomes, a mutation operator is considered in order to avoid GAs getting trapped on a local optimum. This usually happens when most of the individuals in one certain generation are very similar.

In this step, each gene of all individuals is checked according to a predefined mutation rate ρ_m. If a random-generated probability is less than ρ_m, the current gene is mutated by another random value within the considered range. A simple mutation example is given in Fig. 3.9, where the value of the third gene is mutated from 0 to 1.

Crossover point

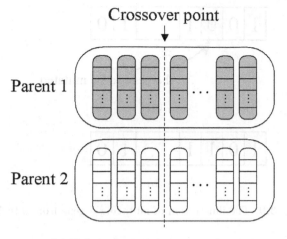

(a) Before the crossover procedure

Crossover point

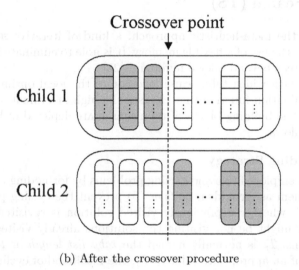

(b) After the crossover procedure

Fig. 3.8. The illustration of the crossover procedure

As in the natural world, ρ_m usually is low. However, it may be varied in different cases. It is usually determined according to experience and knowledge.

3.4.7 Convergence Test

In the GA training procedure, the number of iterations, a predetermined number t, is used. If the number of iterations is met, the training procedure is stopped. The individual with the best performance is the trained output result. The given fitness function can have either the maximum or minimum output.

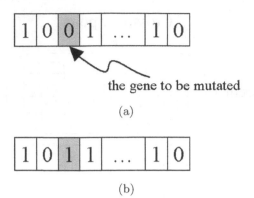

the gene to be mutated

(a)

(b)

Fig. 3.9. The illustration of mutating a gene from 0 to 1

3.5   Tabu Search (TS)

Tabu search is the meta-heuristic approach, a kind of iterative search, and is characterized by the use of a flexible memory. It is able to eliminate local minima and to search areas beyond a local minimum [25].

The process with which tabu search overcomes the local optima problem is based on an evaluation function that chooses the highest evaluation solution at each iteration. The building blocks of tabu search are depicted in Fig. 3.10 and are stated as follows.

3.5.1   Forbidding Strategy

This strategy is employed to avoid cycling problems by forbidding certain moves or classifying them as forbidden, or *tabu*. To prevent the cycling problem, it is sucient to check whether a previously visited solution is revisited or not. An alternative way might be not visiting the solutions already visited during the last T_S iterations. T_S is normally named the *tabu list length* or *tabu list size*. With the help of an appropriate value of T_S, the likelihood of cycling effectively vanishes.

3.5.2   Aspiration Criteria and Tabu Restrictions

An aspiration criterion is applied to make a tabu solution free from the forbidden state if this solution is of sucient quality and can prevent cycling. A solution is acceptable if the tabu restrictions are satisfied. However, a tabu solution is also assumed acceptable if an aspiration criterion applies regardless of the tabu status. We also make use of tabu restrictions to avoid repetitions rather than reversals. A tabu restriction is typically activated only in the case where its attributes occurred within a limited number of iterations prior to the present iteration, or occurred with a certain frequency over a larger number of iterations. Finally, the appropriate use of aspiration criteria can be very significant for enabling a tabu search to achieve its best performance.

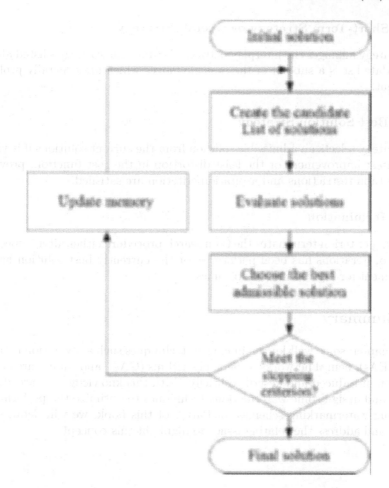

Fig. 3.10. The block diagrams of the TS training procedure

3.5.3   Freeing Strategy

The freeing strategy is taken into account to decide what exits the tabu list. This strategy deletes the tabu restrictions of the solutions so that they can be reconsidered in further steps of the search. The attributes of a tabu solution remain on the tabu list for the duration of T_S iterations.

3.5.4   Intermediate and Long-Term Learning Strategies

These strategies were implemented with intermediate and long-term memory functions. Their operations are recording good features of a selected number of moves generated during the execution of the algorithm.

3.5.5   Short-Term Strategy or Overall Strategy

This strategy manages the interplay between the different strategies listed above. A candidate list is a sub-list of the possible moves which are generally problem dependent.

3.5.6   Best Solution Strategy

This strategy selects an admissible solution from the current solutions if it yields the greatest improvement or the least distortion in the cost function, provided that the tabu restrictions and aspiration criterion are satisfied.

3.5.7   Termination

A stopping criterion terminates the tabu search procedure either after a specified number of iterations has been performed, or the currently best solution has no improvement for a given number of times.

3.6   Summary

In this chapter, some well-known intelligent techniques such as evolutionary algorithms (EAs), neural networks, genetic algorithms (GAs), and tabu search (TS), have been introduced and described briefly. With this knowledge, we are able to proceed and design some watermarking techniques to optimize the performance of existing watermarking schemes. In Part 2 of this book, we will detail some systems and address the relating issues to highlight this concept.

Part II
Intelligent Watermarking

Chapter 4
Spatial-Based Watermarking Schemes and Pixel Selection

Spatial domain based watermarking is one of the fundamental techniques at the beginning of digital watermarking. Generally speaking, spatial domain based watermarking schemes possess the advantages, such as easy implementation and low complexity, than other domain based watermarking schemes. But also, they posses the shortages like weak robustness and non-practical usage than other domain based watermarking schemes.

In this chapter, we concentrate on spatial domain based watermarking techniques and the procedure that improves their performance. We begin with the introduction of this area in the first section, then follow by illustrating two spatial-based watermarking schemes in the second section. The first scheme is the last-significant-bit (LSB) modification method, which is a well-known classical scheme. The second one is a pixel surrounding method, which provides better robustness. Experimental results and comparison are also given in this section. In the third section, a training procedure called genetic pixel selection (GPS) is introduced. It takes the considered methods of attack into account, and employs genetic algorithms to find a better way to select a number of pixels from the cover image to carry the watermark signal. Comparisons of the simulation results with and without the GPS procedure are addressed in the fourth section. Finally, the last section summarises this chapter.

4.1 Introduction

Embedding watermarks into the spatial domain components of the cover images is a straightforward method. It is one of the fundamental schemes used since the beginning of digital watermarking in 1993 [67]. Usually the spatial-based watermarking schemes select a number of pixels from the cover image, and modify the luminance values of these pixels selected according to the watermark bits to be embedded [67] [81] [82]. The image containing these modified pixels therefore now carries the information of the watermark. To extract (or detect) the watermark embedded, usually the same pixels used in the embedding procedure should be selected from the watermarked image firstly. Then, according

F.-H. Wang, J.-S. Pan, and L.C. Jain: Innovations in Dig. Watermark. Tech., SCI 232, pp. 47–62.
springerlink.com

to the strategy used, the bit carried within each pixel can be determined (or detected). By collecting all the bits extracted (or all the results detected), the hidden watermark (or whether the image contains the considered watermark) can be obtained. A well-known classic spatial-based watermarking method is the last-significant-bits (LSB) modification scheme [81]. It selects a number of pixels from the cover image, and modifies their luminance to carry watermark bits. Section 4.2.1 describes its details.

The advantages of spatial-based watermarking schemes are low complexity and easy implementation. However, they also have the weakness that they are not robust against common methods of attack [91], [88], [53]. Due to this, most of the fragile watermarking schemes are based on spatial domain (Sect. 2.2.4). Many methods have been proposed to improve the drawback of weak robustness. Researchers believe that, by the use of channel coding such as the BCH (Bose-Chaudhuri-Hochquenghem) block codes, the performance of spatial-based watermarking schemes can be improved [33]. In [92], the authors presented a scheme having stronger robustness. Like the general scheme mentioned, their scheme also selects the considered number of pixels from the cover image firstly. Then, for each pixel selected, the mean value of its neighbor pixels are calculated. This mean value is referred to modify the pixel selected. The robustness of their scheme is better, which indicates this scheme is more suitable for practical use. Moveover, the scheme also provides a parameter to control the balance between imperceptibility and robustness. Users of this system therefore can decide to have better imperceptibility or better robustness. We illustrate this scheme in Sect. 4.2.2.

4.2   Watermarking Schemes

In this section, the well-known traditional spatial-based watermarking scheme, the LSB method, is introduced first to give readers the general ideas of digital image watermarking. Then, the pixel surrounding method, which improves the shortage of poor robustness proposed in [92], is detailed. Simulation results and comparison given will demonstrate their performance.

4.2.1   LSB Method

In this section, the most traditional spatial-based watermarking scheme, the LSB modification method [81] [82], is illustrated. Figure 4.1 shows its block diagrams.

Let $\mathbf{X}$ denote a cover image which is a set of $M \times H$ pixels. $\mathbf{W}$ denotes a watermark which is a sequence of L bits $\{w_1, w_2, ..., w_L\}$. $\mathbf{K}$ is a key which consists of a set of L integers $\{k_1, k_2, ..., k_L\}$. Here for any integers i and j, where $1 \leq i \leq L$ and $1 \leq j \leq L$, $1 \leq k_i \leq M \times H$ and $k_i \neq k_j$ if $i \neq j$. The user key $\mathbf{K}$ defines which pixels of $\mathbf{X}$ will be selected for carrying watermark bits. That is, k_1 indicates the position of the first pixel selected in $\mathbf{X}$, k_2 indicates the position of the second pixel selected in $\mathbf{X}$, and so on.

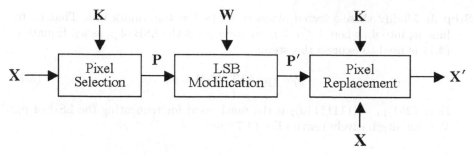

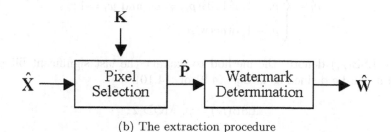

(a) The embedding procedure

(b) The extraction procedure

Fig. 4.1. Block diagrams of the LSB modification scheme

A. Embedding Procedure

Step 1: Generate two position sequences, **A** and **B**, according to the user key **K** and the width M of the cover image **X**. We have:

$$a_i = k_i \ \text{MOD} \ M. \tag{4.1}$$

$$b_i = k_i \ \text{DIV} \ M. \tag{4.2}$$

$$\mathbf{A} = \bigcup_{i=1}^{L} a_i. \tag{4.3}$$

$$\mathbf{B} = \bigcup_{i=1}^{L} b_i. \tag{4.4}$$

Step 2: Select L pixels from **X** to form a pixel set **P** according to **A** and **B**. This becomes:

$$p_i = X(a_i, b_i), \tag{4.5}$$

where $X(a_i, b_i)$ is the pixel at position (a_i, b_i) of **X**. And

$$\mathbf{P} = \bigcup_{i=1}^{L} p_i. \tag{4.6}$$

Step 3: Modify these selected pixels to carry the watermark bits. That is, to hide w_i into p_i, where $1 \leq i \leq L$, we merely set the LSB of p_i as w_i. Equation (4.7) is used to express this step.

$$p_i' = (p_i \text{ AND } 254) + w_i. \tag{4.7}$$

Here $(254)_{10} = (11111110)_2$ is the mask used for truncating the LSB of p_i. We can alternatively rewrite Eq. (4.7) as:

$$p_i' = \begin{cases} p_i, & \text{if LSB}(p_i) = w_i\,; \\ p_i + 1, & \text{if LSB}(p_i) \neq w_i \text{ and } w_i = 1\,; \\ p_i - 1, & \text{otherwise}. \end{cases} \tag{4.8}$$

Here $\text{LSB}(p_i)$ denotes the method to extract the last significant bit of p_i, and it can be defined as in Eqs. (4.9) or (4.10):

$$\text{LSB}(p_i) = p_i \text{ MOD } 2, \tag{4.9}$$

$$\text{LSB}(p_i) = p_i \text{ AND } 1. \tag{4.10}$$

Afterwards, all the modified pixels are collected as the set $\mathbf{P}'$.

$$\mathbf{P}' = \bigcup_{i=1}^{L} p_i'. \tag{4.11}$$

Step 4: Return the modified pixels to their original positions according to $\mathbf{A}$ and $\mathbf{B}$. That is:

$$X(a_i, b_i) = p_i'. \tag{4.12}$$

$$X'(a_i, b_i) = X(a_i, b_i). \tag{4.13}$$

Finally, the watermarked image $\mathbf{X}'$ is obtained.

The elements of $\mathbf{K}$ can be defined by the users or generated randomly. In some cases, it is preferred to use the random-number generator seed to obtain the sequence $\mathbf{K}$, or the sequences $\mathbf{A}$ and $\mathbf{B}$.

B. Extraction Procedure

To extract the hidden watermark from a watermarked image $\hat{\mathbf{X}}$, which may contain natural noise or artificial modification, is not difficult. The steps listed below can be used:

Step 1: Select L pixels from $\hat{\mathbf{X}}$ according to the user key $\mathbf{K}$. Here Steps 1 and 2 of the embedding procedure are repeated.

Step 2: According to the selected pixels, which are known as $\hat{\mathbf{P}}$, the hidden watermark bits can be established by using either Eq. (4.9) or Eq. (4.10):

$$\hat{w}_i = \text{LSB}(\hat{p}_i)\,, \tag{4.14}$$

where $\hat{w}_i$ and $\hat{p}_i$ are the i-th extracted bit and the i-th selected pixel of $\hat{\mathbf{P}}$, respectively.

Step 3: Assemble all the extracted bits $\{\hat{w}_1, \hat{w}_2, ..., \hat{w}_L\}$ to form a recovered watermark $\hat{\mathbf{W}}$.

4.2.2 Pixel Surrounding Method

In this section, another spatial domain based watermarking scheme proposed by Wang et al. [94] is introduced. This scheme provides more advantages such as: good imperceptibility, stronger robustness against the JPEG compression, easy implementation, and a parameter to control the balance between imperceptibility and robustness. Figure 4.2 illustrates the block diagrams of this scheme. Here symbols $\mathbf{X}$, $\mathbf{W}$, and $\mathbf{K}$ have been defined in Sect. 4.2.1.

A. Embedding Procedure

Step 1: Select L pixels from the given cover image $\mathbf{X}$ according to the user key $\mathbf{K}$, as described in Sect. 4.2.1.

Step 2: For each of the pixels selected, calculate the mean values of their surrounding pixels.

$$\mu = \tfrac{1}{8} \begin{bmatrix} 1 & 1 & 1 \\ 1 & 0 & 1 \\ 1 & 1 & 1 \end{bmatrix}, \tag{4.15}$$

where μ is the mean value calculated for current pixel selected. We collect all the mean values calculated as the set $\mathbf{M}$.

$$\mathbf{M} = \bigcup_{i=1}^{L} \mu_i\,. \tag{4.16}$$

Step 3: Modify the selected pixels to contain the watermark bits by referring to a given threshold δ using the equation below. This hides $\mathbf{W}$ in $\mathbf{P}$.

$$p'_i = \begin{cases} \mu_i - \delta, & \text{if } w_i = 0\,; \\ \mu_i + \delta, & \text{otherwise}\,. \end{cases} \tag{4.17}$$

The modified pixels are then formed the output pixel set $\mathbf{P}'$:

$$\mathbf{P}' = \bigcup_{i=1}^{L} p'_i\,. \tag{4.18}$$

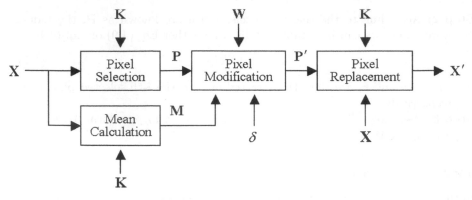

(a) The embedding procedure

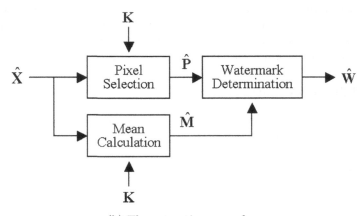

(b) The extraction procedure

Fig. 4.2. Block diagrams of Wang et al.'s spatial-based watermarking scheme

Step 4: Generate the watermarked image $\mathbf{X}'$ by inserting the modified pixels back to their original positions as specified by $\mathbf{K}$.

In the embedding procedure above, the value of δ controls the balance between imperceptibility and robustness. That is, a smaller δ means the quality of the watermarked image will be better but the robustness is weaker.

B. Extraction Procedure

The hidden watermark can be extracted from the watermarked image by using the steps below. Here the original non-watermarked image is not required.

Step 1: Select L pixels from the given watermarked image, $\hat{\mathbf{X}}$, according to the user key, $\mathbf{K}$. This step is the same as the first step in the embedding procedure.

Step 2: Calculate the mean values for these selected pixels by using Eq. (4.15). This is described in Step 2 of the embedding procedure. $\hat{\mathbf{M}}$ is then used to denote all the calculated mean values $\{\hat{\mu}_1, \hat{\mu}_2, ..., \hat{\mu}_L\}$.

Step 3: Determine the bits hidden in the selected pixels to form the recovered watermark $\hat{\mathbf{W}}$. This is done by applying:

$$\hat{w}_i = \begin{cases} 0, \text{ if } \hat{p}_i < \hat{\mu}_i; \\ 1, \text{ otherwise.} \end{cases} \tag{4.19}$$

$$\hat{\mathbf{W}} = \bigcup_{i=1}^{L} \hat{w}_i. \tag{4.20}$$

4.2.3   Simulation Results

To demonstrate the performance of the mentioned schemes, the well-known test images LENA (Fig. 1.2) and PEPPERS (Fig. 1.3) were used as the cover images. The image of ROSE (Fig. 1.1(a)) was used as the watermark. The specification of these images are 512×512 pixels in gray-level for either of the cover image and 128×128 pixels in bi-level for the watermark.

To test the robustness of the proposed watermarking system, the JPEG compression using different quality factors (QF) was employed to attack the watermarked images. The reason for using the JPEG compression as the attacking agent is due to the popularity of transmitting and storing digital images in JPEG format. The peak-signal-to-noise ratio (PSNR, see Eq. (2.4)) and the bit correct rate (BCR, see Eq. (2.7)) were applied to evaluate imperceptibility and robustness respectively.

A. Results of the LSB Method

For the LSB method introduced in Sect. 4.2.1, except using the last bit, which is the 8th bit, of each selected pixel to carry the watermark bit, we also used the 5th, 6th, and the 7th bit respectively to hide the watermark bit. We afterwards employed the JPEG compression with a QF=90% to attack the watermarked images. The imperceptibility test results, which are the PSNR values, and the robustness results, which are the BCR values, are given in Table 4.1. The LSB method has very poor robustness against common attack schemes, as described in Sect. 2.2.3. We therefore only display one extracted watermark in Fig. 4.3 as the example, where the watermark bits are embedded in the 6th bits of the selected pixels.

B. Results of the Pixel Surrounding Method

When using the image of LENA as the cover image, Table 4.2 summarises the performance of the pixels surrounding method and Fig. 4.4 displays the extracted watermarks. When using the image of PEPPERS as the cover image, Table 4.3 and Fig. 4.5 show the performance.

Table 4.1. The imperceptibility test results (PSNR) and the robustness test results (BCR) against the JPEG attack when the image of LENA and a QF=90% are used

Position of the embedded bit	PSNR (dB)	BCR (%)
5th	45.09	61.03
6th	51.14	53.45
7th	57.15	50.61
8th	63.21	50.38

Fig. 4.3. The watermark extracted from the watermarked image which is attacked by the JPEG compression when the image of LENA and a QF=90% are used

Table 4.2. PSNR values and BCR values under the JPEG attacks when the image of LENA and different values of δ are used.

δ	PSNR (dB)	BCR (%)		
		QF=70%	QF=80%	QF=90%
5	42.43	60.74	65.23	78.28
10	38.89	72.59	80.19	95.20
15	36.04	82.50	90.72	99.20

Table 4.3. PSNR values and BCR values under the JPEG attacks when the image of PEPPERS and different values of δ are used.

δ	PSNR (dB)	BCR (%)		
		QF=70%	QF=80%	QF=90%
5	42.21	61.85	66.04	80.07
10	38.79	73.55	80.76	95.89
15	35.99	83.42	90.94	99.41

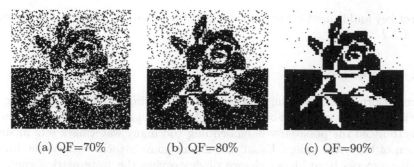

(a) QF=70%   (b) QF=80%   (c) QF=90%

Fig. 4.4. The extracted watermarks under the JPEG attack when the image of LENA and a $\delta = 15$ are used

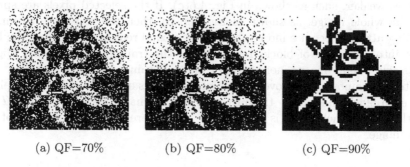

(a) QF=70%   (b) QF=80%   (c) QF=90%

Fig. 4.5. The extracted watermarks under the JPEG attack when the image of PEPPERS and a $\delta = 15$ are used

4.2.4 Comparison and Discussion

Generally speaking, compared with the transform-based watermarking techniques, spatial-based watermarking schemes usually have weaker robustness against common attack schemes. Experimental results of the LSB method presented in Sect. 4.2.3 have shown this. However, Wang et al.'s watermarking scheme provides better robustness against the JPEG compression. It implies that the later scheme is more suitable for practical use. This is because digital images are now mostly stored and transferred in the JPEG format.

4.3 Genetic Pixel Selection (GPS)

As mentioned in Sect. 4.2, the first step for spatial-based watermarking schemes usually is selecting the needed pixels from the cover image. It is also noted that spatial-based watermarking schemes generally have weak robustness against common attacks. Thus, the ideas of designing a training procedure, which uses GAs to select suitable pixels and takes the considered attacks into account, for spatial-based watermarking schemes to improve their performance including imperceptibility and robustness [94] is introduced. An object function is defined and GA is employed to optimize it. After that, the trained result is regarded as a secret key and is used in the spatial-based watermarking scheme introduced in Sect. 4.2.2.

4.3.1 Preprocessing

Before performing the training process, a preprocessing step for decomposing the cover image into several non-overlapping blocks is suggested. The first reason for this is to avoid the possibility of embedding too many watermark bits in single location of the cover image. Usually, it means some attacks such as the image-cropping process have higher chance of destroying the watermark signal. For instance, as shown in Fig. 4.6, when some selected pixels are gathered too close such as in Fig. 4.6(a), the robustness against image-cropping attacks sometimes becomes weaker, such as shown in Fig. 4.6(c). If the selected pixels are spread over the whole image, as shown in Fig. 4.6(b), then the resistance to image-cropping attacks becomes more robust. The second reason for dividing the host image into a number of blocks is to shorten the training time of GAs. Bigger blocks size usually means more search time to obtain a better pixel set.

After decomposing the cover image into T blocks of size D, we select b pixels from each block and modify them to carry b watermark bits. That is, $L = T \times b$ and the size of each block is $D = \frac{M \times H}{T} > b$, where $M \times H$ is the size of the cover image.

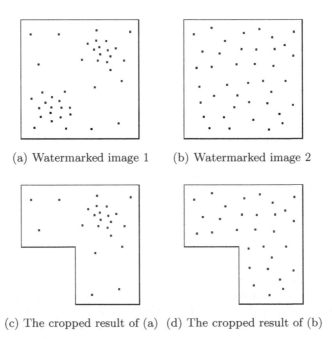

(a) Watermarked image 1   (b) Watermarked image 2

(c) The cropped result of (a) (d) The cropped result of (b)

Fig. 4.6. The effect on the distribution of the pixels selected under image-cropping attacks. The dots in the figures denote the pixels selected.

4.3.2 Training Scheme

The definitions of the symbols used in Sects. 4.2.1 and 4.2.2 are now repeated. Let $\mathbf{X}$ be a cover image which is a set of $M \times H$ pixels, $\mathbf{W}$ be a watermark which is a set of L bits, and $\mathbf{K}$ be a user key which is a set of L integers $\{k_1, k_2, ..., k_L\}$. Here for any integers $i : 1 \leq i \leq L$ and $j : 1 \leq j \leq L$, $1 \leq k_i \leq M \times H$ and $k_i \neq k_j$ if $i \neq j$. The user key defines which pixels of $\mathbf{X}$ will be selected for carrying watermark bits. That is, k_1 indicates the position of the first pixel selected in $\mathbf{X}$, k_2 indicates the position of the second pixel selected in $\mathbf{X}$, and so on. The steps below are then used as the GPS procedure.

A. Steps

Step 1: Generate S keys $\{\mathbf{K}_1, \mathbf{K}_2, ..., \mathbf{K}_S\}$ randomly as the initial GA individuals.

Step 2: Embed the watermark into the cover image using these keys. (S watermarked images will be obtained afterwards.)

Step 3: Apply the considered attack scheme to attack these watermarked images one by one. (S attacked images can be then obtained.)

Step 4: Extract S watermarks from these attacked images one by one.

Step 5: Evaluate the performance for all individuals according to the imperceptibility and robustness results.

Step 6: Select the individuals with better performance.

Step 7: Stop the training procedure if the considered GA iteration t is met.

Step 8: Execute the crossover procedure to generate new individuals for next generation.

Step 9: Perform the mutation procedure on the new individuals.

Step 10: Go to Step 2.

Details of the above steps are given in the following sections.

B. Details of the GPS Procedure

The goal of the GPS procedure is to select a better set of pixels for embedding, the user key $\mathbf{K}$ therefore is regarded as an individual in Step 1 of the GPS procedure listed above. Each element of $\mathbf{K}$ is regarded as a gene. The needed S individuals are generated randomly according to the definition of $\mathbf{K}$.

In Step 2, the spatial-based watermarking scheme mentioned in Sect. 4.2.2B is employed. Each generated key is used to select the needed number of pixels from the cover image $\mathbf{X}$, and the watermark bits are then embedded within these pixels. When repeating this embedding step S times, where in each time the used key for pixel selection is different, S watermarked images $\{\mathbf{X}'_1, \mathbf{X}'_2, ..., \mathbf{X}'_S\}$ can be obtained. The considered attack scheme is applied to attack the watermarked images one by one and to generate S attacked images $\{\hat{\mathbf{X}}_1, \hat{\mathbf{X}}_2, ..., \hat{\mathbf{X}}_S\}$. From each attacked image, a watermark can be extracted using the watermark extraction procedure introduced in Sect. 4.2.2. We then have S extracted watermarks $\{\mathbf{W}'_1, \mathbf{W}'_2, ..., \mathbf{W}'_S\}$.

To evaluate the performance as regards imperceptibility and robustness for $\{\mathbf{K}_1, \mathbf{K}_2, ..., \mathbf{K}_S\}$, a fitness function defined as Eq. (4.21) is used:

$$f_i = f_I(\mathbf{X}, \mathbf{X}'_i) + \lambda \times f_R(\mathbf{W}_i, \mathbf{W}'_i), \; i = 1, 2, ..., S, \qquad (4.21)$$

where f_i denotes the performance of the i-th individual $\mathbf{K}_i$, f_I is the imperceptibility evaluating function (e.g. PSNR), f_R is the robustness evaluating function (e.g. BCR), and λ is a parameter used for controlling the balance between f_I and f_R. In Step 5, the fitness scores of the individuals are calculated to measure their performance.

In Step 6, according to these scores, the probabilities for being selected as parents in the crossover step are assigned. Usually, the higher fitness score the individual has, the higher probability of becoming a parent the individual is assigned. According to the probabilities, the crossover process selects two parents to produce two children for next generation. After generating S new individuals, Step 9 is performed to mutate the genes of the new individuals according to a given mutation rate ρ_m. That is, for the j-th ($1 \leq j \leq S$) individual $\mathbf{K}_j$ and its i-th ($1 \leq i \leq L$) gene k_i^j, if the probability generated by the random-number generator is less than ρ_m, then k_i^j is mutated as k', where $0 \leq k' < M \times H$ and for any integer $l : 1 \leq l \leq L$, $k' \neq k_l^j$. If the preprocessing procedure has been applied beforehand, then $0 \leq k' < D$ and for any integer $l : 1 \leq l < b$, $k' \neq k_l^j$.

The above steps are repeated until the iteration t is satisfied. Finally, the individual with the best fitness score in the final generation is the output trained result.

4.3.3 Simulation Results

In the experiments, the spatial-based watermarking scheme presented in Sect. 4.2.2 was employed in the GPS procedure. For the purpose of comparing with the performance with and without the GPS procedure, the same data and settings used in Sect. 4.2.3 were also used here. That is, the image of LENA (Fig. 1.2) and the image of PEPPERS (Fig. 1.3) were regarded as the cover images, and the image of ROSE (Fig. 1.1(a)) was regarded as the watermark. The sizes of the watermark and either cover image are 128×128 pixels in bi-level and 512×512 pixels in gray-level respectively. The JPEG compression with different quality factors (QF) was used to attack the watermarked images. The other settings used are: $S = 10$, $t = 100$, and $\lambda = 100 \times \text{PSNR}(\mathbf{X}, \mathbf{X}')$.

When the image of LENA was used as the cover image, Table 4.4 lists the performance with the GPS procedure. Figure 4.7 demonstrates the watermarks extracted from the attacked images when a $\delta = 15$ was used. When using the image of PEPPERS as the cover image, Table 4.5 lists the performance with the GPS procedure. Figure 4.8 displays the watermarks extracted from the attacked images when a $\delta = 15$ was used.

Table 4.4. Performance with the GPS procedure when LENA is used

δ	PSNR (dB)	BCR (%)		
		QF=70%	QF=80%	QF=90%
5	51.90	78.91	80.08	88.49
10	46.81	88.43	91.70	98.21
15	42.96	93.95	96.41	99.73

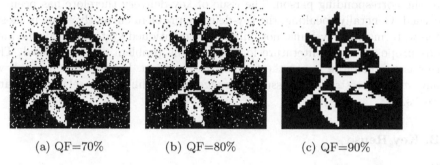

(a) QF=70%   (b) QF=80%   (c) QF=90%

Fig. 4.7. The extracted watermarks under the JPEG attacks when using LENA as the cover image and δ = 15.

Table 4.5. Performance with the GPS procedure when PEPPERS is used

δ	PSNR (dB)	BCR (%)		
		QF=70%	QF=80%	QF=90%
5	51.66	74.78	75.51	85.11
10	47.09	85.75	89.12	97.96
15	43.44	92.16	94.98	99.71

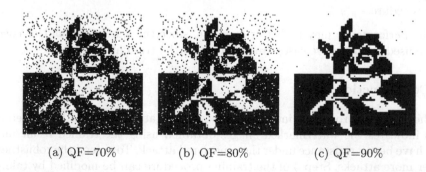

(a) QF=70%   (b) QF=80%   (c) QF=90%

Fig. 4.8. The extracted watermarks under the JPEG attacks when using PEPPERS as the cover image and δ = 15

4.3.4 Discussion

Some issues associated with the GPS procedure are discussed here.

A. Key Delivery

As we have known, the training procedure finds a better way to select a number of pixels from the cover image. Then, the trained result is regarded as a key used in the considered watermarking system. In the real world, sometimes users of a watermarking system may have the problems in delivering the key to the corresponding person. The issue of key-delivery therefore has been concerned. Generally speaking, the key-delivery problem currently exists in most of the watermarking systems, not only in the mentioned systems. Also, the methods proposed in the literature for solving this problem can be applied. Thus, this issue therefore is not the focus of this section. Readers of this book who have the interests in this issue are suggested reading some relating literature works.

B. Key Reuse

Key-reuse is another issue encountered and concerned in many systems. For the systems introduced, basically the generated key is associated with the cover image and the watermark used in the training procedure. It means the performance can be improved only when using this key to embed the considered watermark within the considered cover image. In other words, in the below cases, it results in varying performance:

(1) Using the key to embed the considered watermark into another cover image, which is unknown to the training procedure.
(2) Using the key to embed a watermark, which is unknown to the training procedure, into the considered cover image.
(3) Using the key to embed a watermark, which is unknown to the training procedure, into another cover image, which is also unknown to the training procedure.

The performance may be better or worse, since the watermark and/or cover image used in the above cases are unknown to the training procedure.

C. Cost Function

In the training procedure introduced above, one attack is considered merely. This indicates the watermarked images obtained using the trained result may only have better resistance under the considered attack. To extend the robustness under more attacks, Step 3 of the training procedure can be modified by taking more attacks into account, such as Fig. 4.9.

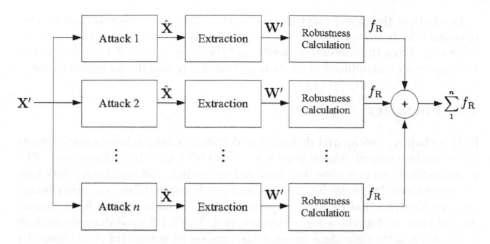

Fig. 4.9. Considering more attacks in the training procedure

In Fig. 4.9, after extracting a watermark out from each attacked image, the robustness score can be calculated. After that, the average robustness score is calculated and used in Eq. 4.21, which now is modified as:

$$f_i = f_{\mathrm{I}}(\mathbf{X}, \mathbf{X}'_i) + \lambda \times \frac{1}{n} \sum_{j=1}^{n} f_{\mathrm{R}}(\mathbf{W}_{i,j}, \mathbf{W}'_{i,j}), \; i = 1, 2, ..., S. \qquad (4.22)$$

It should be noted that:

(1) Taking every kind of attacks into account is not practical.
(2) Some attacks are not suitable to be employed in the training procedure. For instance, what will happen while employing the random noise as the attacking function?
(3) Employing some attacks which have exclusive characteristics may result in bad output. For example, what will happen while employing low-pass filtering as the first attack and high-pass filtering as the second attack?

Answers of above questions are left to the reader.

4.4 Comparison

Comparing Table 4.2 with Table 4.4, and Table 4.3 with Table 4.5, it is obvious that the GPS procedure improves the performance both in imperceptibility and in robustness. Although the GPS training procedure requires time to generate a better key for embedding and extraction, this training duty can be achieved off-line. In other words, if a system does not require the online training, the proposed GPS scheme can be employed upon the watermarking system to obtain better performance.

In addition, the object function used in the GPS procedure also shows a way to control the balance between imperceptibility and robustness by changing the value of λ. From the simulation results and the comparisons, it is clear that the GPS procedure introduced plays its role successfully within our expectations.

4.5   Summary

In this chapter, two spatial domain based watermarking schemes and a training procedure named genetic pixel selection (GPS) have been illustrated. The watermarking schemes show the readers how to hide the watermark bits into the cover image by modifying the values of pixels selected from the cover image directly. Experimental results, comparisons, and discussions given have demonstrated their performance and characteristics. The GPS procedure shows how to employ genetic algorithms to select the considered number of pixels from the cover image for watermarking. Experimental results given have demonstrated that by introducing the GPS procedure, the mentioned spatial domain based watermarking scheme can provide better performance. This echoes the concept we mentioned in the first chapter: By introducing intelligent techniques into watermarking schemes, the performance of them may be improved.

Chapter 5
Discrete Cosine Transform Based Watermarking Scheme and Band Selection

Compared with spatial domain based watermarking techniques, transform domain based watermarking techniques have become the main stream of this research area for a long time, since transform domain based watermarking schemes can provide not only good watermarked image quality, but also stronger robustness under general attacks or noise affection. In this chapter, our focus is shifting to the transform domain based watermarking scheme, where a watermarking scheme based on the most popular discrete cosine transform (DCT) is presented. The DCT-based scheme first transforms the cover image into frequency domain. It then selects a number of DCT bands according to the user-specified key and modifies these bands to carry the watermark bits. To have better coding results, the concept of introducing intelligent techniques into the watermarking scheme is employed again. Here a training procedure named genetic band selection (GBS) is illustrated. It employs the genetic algorithm (GA) to select a group of suitable DCT bands for watermarking. The trained result is then used in the mentioned DCT-based watermarking scheme. With the trained result of the GBS procedure, we expect the original watermarking scheme could have better performance.

We begin with the introduction of the general background in Sect. 5.1, then detail the DCT-based watermarking method in Sect. 5.2. Experimental results, comparisons, and discussions are also included in this section. In Sect. 5.3, the GBS procedure is explained. Here except the demonstration of the simulation results, comparisons and discussions are also given here to highlight its performance. Finally, Sect. 5.4 summarises this chapter.

5.1 Discrete Consine Transform (DCT)

Let $\mathbf{x}$ be an image containing $M \times N$ pixels therein, $x[i][j]$, where $i = 0, 1, ..., M - 1$ and $j = 0, 1, ..., N - 1$, be the pixel at position (i, j) of $\mathbf{x}$, and $y[u][v]$, where

F.-H. Wang, J.-S. Pan, and L.C. Jain: Innovations in Dig. Watermark. Tech., SCI 232, pp. 63–82.
springerlink.com   © Springer-Verlag Berlin Heidelberg 2009

$u = 0, 1, ..., M-1$ and $v = 0, 1, ..., N-1$, be the transformed coefficient of $x[i][j]$. We have:

$$y[u][v] = \frac{2c(u)c(v)}{\sqrt{MN}} \sum_{i=0}^{M-1} \sum_{j=0}^{N-1} x[i][j] \cos \left[\frac{(2i+1)u\pi}{2M} \right] \cos \left[\frac{(2j+1)v\pi}{2N} \right]. \quad (5.1)$$

$$x[i][j] = c(u)c(v) \sum_{u=0}^{M-1} \sum_{v=0}^{N-1} y[u][v] \cos \left[\frac{(2i+1)u\pi}{2M} \right] \cos \left[\frac{(2j+1)v\pi}{2N} \right]. \quad (5.2)$$

$$c(k) = \begin{cases} \frac{1}{\sqrt{2}}, & \text{if } k = 0 \, ; \\ 1, & \text{otherwise} \, . \end{cases} \quad (5.3)$$

Thus, all the transformed coefficients can be collected as a set $\mathbf{y}$:

$$\mathbf{y} = \bigcup_{u=0}^{M-1} \bigcup_{v=0}^{N-1} y[u][v]. \quad (5.4)$$

We summarise the above steps as Eq. (5.5), and the inverse DCT procedure as Eq. (5.6):

$$\mathbf{y} = \mathrm{DCT}(\mathbf{x}) \, . \quad (5.5)$$

$$\mathbf{x} = \mathrm{DCT}^{-1}(\mathbf{y}) \, . \quad (5.6)$$

Generally speaking, to transform a large size of $\mathbf{x}$ to $\mathbf{y}$ is not practical. Decompositing $\mathbf{x}$ into a number of non-overlapping sub-images, and apply DCT to each of them therefore is applied in the literature. For example, many people decompose the input image into blocks with size 8×8. Figure 5.1 presents one example to show the relationship between $\mathbf{x}$, the spatial domain pixels, and $\mathbf{y}$, the DCT coefficients. Here the DCT coefficients are arranged in the zigzag ordered.

Also, in the DCT coefficients obtained, the first DCT coefficient, such as $y[0]$ in Fig. 5.1, is regarded as the DC value of the block, and the others are regarded as the AC values. As human eyes are much more sensitive to low-frequency bands, we therefore can eliminate some high-frequency AC bands to achieve the goal of data compression.

5.2 DCT-Based Watermarking Scheme

Transform domain based watermarking techniques are the most popular ones in the research area of digital watermarking, and many researchers have already proposed a variety of schemes and applications. Among them, discrete

$x[0][0]$	$x[0][1]$	$x[0][2]$	$x[0][3]$	$x[0][4]$	$x[0][5]$	$x[0][6]$	$x[0][7]$
$x[1][0]$	$x[1][1]$	$x[1][2]$	$x[1][3]$	$x[1][4]$	$x[1][5]$	$x[1][6]$	$x[1][7]$
$x[2][0]$	$x[2][1]$	$x[2][2]$	$x[2][3]$	$x[2][4]$	$x[2][5]$	$x[2][6]$	$x[2][7]$
$x[3][0]$	$x[3][1]$	$x[3][2]$	$x[3][3]$	$x[3][4]$	$x[3][5]$	$x[3][6]$	$x[3][7]$
$x[4][0]$	$x[4][1]$	$x[4][2]$	$x[4][3]$	$x[4][4]$	$x[4][5]$	$x[4][6]$	$x[4][7]$
$x[5][0]$	$x[5][1]$	$x[5][2]$	$x[5][3]$	$x[5][4]$	$x[5][5]$	$x[5][6]$	$x[5][7]$
$x[6][0]$	$x[6][1]$	$x[6][2]$	$x[6][3]$	$x[6][4]$	$x[6][5]$	$x[6][6]$	$x[6][7]$
$x[7][0]$	$x[7][1]$	$x[7][2]$	$x[7][3]$	$x[7][4]$	$x[7][5]$	$x[7][6]$	$x[7][7]$

$$\Downarrow \ 8 \times 8 \ \text{DCT}$$

$y[0]$	$y[1]$	$y[5]$	$y[6]$	$y[14]$	$y[15]$	$y[27]$	$y[28]$
$y[2]$	$y[4]$	$y[7]$	$y[13]$	$y[16]$	$y[26]$	$y[29]$	$y[42]$
$y[3]$	$y[8]$	$y[12]$	$y[17]$	$y[25]$	$y[30]$	$y[41]$	$y[43]$
$y[9]$	$y[11]$	$y[18]$	$y[24]$	$y[31]$	$y[40]$	$y[44]$	$y[53]$
$y[10]$	$y[19]$	$y[23]$	$y[32]$	$y[39]$	$y[45]$	$y[52]$	$y[54]$
$y[20]$	$y[22]$	$y[33]$	$y[38]$	$y[46]$	$y[51]$	$y[55]$	$y[60]$
$y[21]$	$y[34]$	$y[37]$	$y[47]$	$y[50]$	$y[56]$	$y[59]$	$y[61]$
$y[35]$	$y[36]$	$y[48]$	$y[49]$	$y[57]$	$y[58]$	$y[62]$	$y[63]$

Fig. 5.1. The illustrations between spatial-domain pixels **x** and the DCT coefficients **y**, where the DCT coefficients are arranged in the zigzag ordered

cosine transform (DCT) [2] and discrete wavelet transform (DWT) [87] are the
most popular transform coding schemes employed in the literature. Here in this
section, a DCT-based watermarking approach proposed by Shieh et al. [77] is
introduced. Details of it are illustrated in the following subsections.

5.2.1   Embedding Procedure

Let the input image be $\mathbf{X}$ with size $M \times H$, and the binary-valued watermark be
$\mathbf{W}$ with size $M_W \times H_W$. Our goal here is to embed $\mathbf{W}$ into $\mathbf{X}$ in the DCT domain
by selecting some DCT coefficients from $\mathbf{X}$ and modifying them to carry the
watermark bits. Figure 5.2 shows the block diagram of the embedding procedure,
and the steps followed are used as the embedding procedure.

Step 1: Decompose $\mathbf{X}$ into T non-overlapping blocks $\{\mathbf{x}_1, \mathbf{x}_2, ..., \mathbf{x}_T\}$ with size
    8×8 pixels, where $T = \frac{M \times H}{8 \times 8}$.

Step 2: Disperse the spatial relationships of $\mathbf{W}$. Doing so helps provide better
    robustness under the cropping attacks. We use $\mathbf{W}_P$ to denote the permuted
    version of $\mathbf{W}$.

Step 3: Decompose $\mathbf{W}_P$ into T non-overlapping sections $\{\mathbf{w}_1, \mathbf{w}_2, ..., \mathbf{w}_T\}$. Here
    the length of each section is $L = \frac{M_W \times H_W}{T}$ bits.

Step 4: Perform the 8×8 block DCT (see Eq. (5.5)) on $\{\mathbf{x}_1, \mathbf{x}_2, ..., \mathbf{x}_T\}$ one by
    one, and get the DCT coefficients $\{\mathbf{y}_1, \mathbf{y}_2, ..., \mathbf{y}_T\}$ respectively.

Step 5: Generate the reference table $\mathbf{R}$ from $\{\mathbf{y}_1, \mathbf{y}_2, ..., \mathbf{y}_T\}$. With this table,
    we are able to modify the necessary DCT coefficients for carrying the water-
    mark bits.

Step 6: Select L coefficients from $\mathbf{y}_i$ $(1 \leq i \leq T)$ according to the user-specified
    key $\mathbf{S}$.

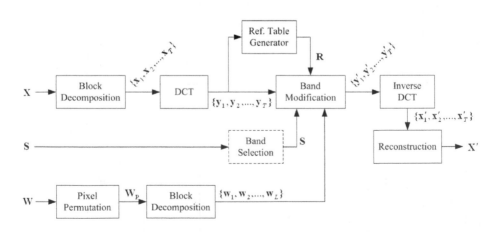

Fig. 5.2. Block diagram of the embedding procedure

Step 7: Modify each selected DCT coefficient according to the corresponding bit contained in $\mathbf{w}_i$, repectively. Here the reference table $\mathbf{R}$ is referred and the modified result is denoted as $\mathbf{y}_i'$.

Step 8: Repeat Steps 6 and 7 until all the DCT blocks have been processed. Afterwards, we have $\{\mathbf{y}_1', \mathbf{y}_2', ..., \mathbf{y}_T'\}$.

Step 9: Apply the inverse DCT (see Eq. (5.6)) on $\{\mathbf{y}_1', \mathbf{y}_2', ..., \mathbf{y}_T'\}$. The reconstructed blocks are denoted as $\{\mathbf{x}_1', \mathbf{x}_2', ..., \mathbf{x}_T'\}$.

Step 10: Finally, piece together $\{\mathbf{x}_1', \mathbf{x}_2', ..., \mathbf{x}_T'\}$ to form a reconstructed image $\mathbf{X}'$. This image is also the watermarked image.

Details of the above procedure are described followed.

A. Pixel Permutation

As the concept mentioned in Sect. 4.3.1, to provide better robustness under the attacks of cropping, a pixel permutation procedure is suggested here. This procedure applies a pseudo-random number generator (PRNG) to disperse the spatial relationships of the watermark. We have:

$$\mathbf{W}_P = \text{Permute}(\mathbf{W}, seed), \qquad (5.7)$$

where *seed* is a pre-determined seed to be used in the pseudo-random number generating system, and $\mathbf{W}_P$ is the permuted version of the input watermark $\mathbf{W}$. Figure 5.3 shows an example of a watermark and the permuted verison of it.

To recover the original watermark $\mathbf{W}$ from its permuted version $\mathbf{W}_P$, the inverse process with the same seed used in the above pixel permution process can be applied. That is:

$$\mathbf{W} = \text{Permute}^{-1}(\mathbf{W}_P, seed). \qquad (5.8)$$

In the embedding procedure illustrated followed, instead of embedding $\mathbf{W}$, we will embed $\mathbf{W}_P$ into the cover image.

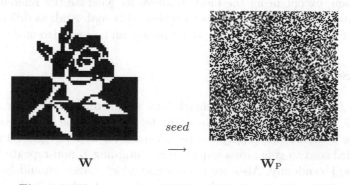

$$\mathbf{W} \qquad \xrightarrow{seed} \qquad \mathbf{W}_P$$

Fig. 5.3. The watermark ROSE and its permuted version

B. Reference Table

In [77], to embed the watermark bits into the cover image in DCT domain, a reference table $\mathbf{R}$ is required. Generally speaking, this reference table is a set of 64 coefficients, which can be defined off-line by the user (thus can be regarded as a secret key) or generated from the used cover image during watermarking. For example, the below method can be applied to generate a reference table from the cover image.

Let $\{\mathbf{y}_1, \mathbf{y}_2, ..., \mathbf{y}_T\}$ be the DCT coefficients of $\{\mathbf{x}_1, \mathbf{x}_2, ..., \mathbf{x}_T\}$ respectively. Here for any $j : 1 \leq j \leq T$, $\mathbf{y}_j$ contains 64 DCT coefficients $\{y_j[0], y_j[1], ..., y_j[63]\}$. We first sum up all the i-th frequences of all DCT blocks:

$$\mu[i] = \sum_{j=1}^{T} |y_j[i]|, \ i = 0, 1, 2, ..., 63. \tag{5.9}$$

Then, after $\{\mu[0], \mu[1], \mu[2], ..., \mu[63]\}$ are calculated, the reference table $\mathbf{R}$ can be generated:

$$R[i] = \begin{cases} \mu[0]/T, \text{ if } \mu[i] = 0; \\ \mu[i]/T, \text{ otherwise.} \end{cases} \tag{5.10}$$

$$\mathbf{R} = \bigcup_{i=0}^{63} R[i]. \tag{5.11}$$

This reference table will be referred while modifying the DCT coefficients for carrying the watermark bits.

In the above method, as it is well known that altering the DC value of a DCT block usually causes large distortion while reconstruction, few people therefore modify DC values to carry watermark bits. This means the first element of $\mathbf{R}$, which is $R[0]$, usually will not be referred. Thus, we can ignore calculating $R[0]$.

In addition, except using the method above to generate the reference table, users can also employ other methods to achieve this goal. And, as different users can define different reference tables, $\mathbf{R}$ therefore can be regarded as a secret key.

C. Band Selection

To embed L ($1 \leq L < 64$) watermark bits within one DCT block, we have to select L bands (or coefficients in otehr terms) from the 63 AC bands. To achieve this goal, we may apply the pseudo-random number generator with a user-specified seed to generate a sequence $\mathbf{S}$ containing L non-repeated integers $\{s_1, s_2, ..., s_L\}$ randomly. Also, we may define which bands should be selected according to our preference or employ other methods introduced in the literature. According to this sequence, the $(s_1 + 1)$-th, $(s_2 + 1)$-th, ..., and $(s_L + 1)$-th bands are selected. Below is an example which demonstrates selecting four DCT bands for watermarking.

Assume that $L = 4$ and $\mathbf{S} = \{13, 19, 24, 25\}$ is a user-specified sequence. For a given DCT block $\mathbf{y}$, according to the content of $\mathbf{S}$, the 14-th, 20-th, 25-th, and the 26-th bands, which are $\{y[13], y[19], y[24], y[25]\}$, are selected, as shown in Fig. 5.4.

$y[0]$	$y[1]$	$y[5]$	$y[6]$	$y[14]$	$y[15]$	$y[27]$	$y[28]$
$y[2]$	$y[4]$	$y[7]$	$y[13]$	$y[16]$	$y[26]$	$y[29]$	$y[42]$
$y[3]$	$y[8]$	$y[12]$	$y[17]$	$y[25]$	$y[30]$	$y[41]$	$y[43]$
$y[9]$	$y[11]$	$y[18]$	$y[24]$	$y[31]$	$y[40]$	$y[44]$	$y[53]$
$y[10]$	$y[19]$	$y[23]$	$y[32]$	$y[39]$	$y[45]$	$y[52]$	$y[54]$
$y[20]$	$y[22]$	$y[33]$	$y[38]$	$y[46]$	$y[51]$	$y[55]$	$y[60]$
$y[21]$	$y[34]$	$y[37]$	$y[47]$	$y[50]$	$y[56]$	$y[59]$	$y[61]$
$y[35]$	$y[36]$	$y[48]$	$y[49]$	$y[57]$	$y[58]$	$y[62]$	$y[63]$

Fig. 5.4. Example of selecting 4 bands, $y[13], y[19], y[24]$, and $y[25]$, from a DCT block

In the example above, since different users may define different sequences, $\mathbf{S}$ therefore can be regarded as a secret key. And, due to there are T DCT blocks $\{\mathbf{y}_1, \mathbf{y}_2, ..., \mathbf{y}_T\}$ to be dealt with, we can define T sub-sequences, that is, $\mathbf{S} = \{\mathbf{s}_1, \mathbf{s}_2, ..., \mathbf{s}_T\}$ and for any $\mathbf{s} \in \mathbf{S}$, $\mathbf{s} = \{s_1, s_2, ..., s_L\}$, to specify the positions of the bands to be picked for $\{\mathbf{y}_1, \mathbf{y}_2, ..., \mathbf{y}_T\}$ respectively, or just simply apply the same sequence, that is, $\mathbf{S} = \{s_1, s_2, ..., s_L\}$ to select the required number of DCT bands from all the DCT blocks. Also, as mentioned previously that altering the

DC value (the first band) of a DCT block usually causes large distortion while reconstruction, we therefore avoid selecting the first band and only select the middle bands.

D. Band Modification

Let $\mathbf{y}$ be a DCT block containing 64 DCT bands $\{y[0], y[1], ..., y[63]\}$, $\mathbf{w}$ be a watermark containing L binary bits $\{w[1], w[2], ..., w[L]\}$, $\mathbf{R}$ be a reference table containing 63 coefficients $\{R[1], R[2], ...R[63]\}$, and $\mathbf{s} = \{s[1], s[2], ..., s[L]\}$ be a sequence indicating which DCT bands of $\mathbf{y}$ should be picked. To embed the binary watermark $\mathbf{w}$ into $\mathbf{y}$, we firstly determine the *polarities* of the bands to be modified for carrying the watermark bits. That is:

$$P[i] = \begin{cases} 1, \text{ if } y[s[i]] \cdot R[s[i]] \geq y[0]; \ i = 1, 2, ..., L; \\ 0, \text{ otherwise.} \end{cases} \tag{5.12}$$

Afterwards, the below equation is used for modifying the selected DCT bands:

$$y'[s[i]] = \begin{cases} y[s[i]], & \text{if } P[i] = w[i]; \\ y[0]/y[s[i]] + 1, & \text{if } P[i] = 0 \text{ and } w[i] = 1; \\ y[0]/y[s[i]] - 1, & \text{if } P[i] = 1 \text{ and } w[i] = 0. \end{cases} \tag{5.13}$$

After all the necessary DCT bands of $\mathbf{y}$ have been modified to carry the L watermark bits, we have the watermarked DCT block $\mathbf{y}'$.

5.2.2   Extraction Procedure

To obtain the embedded watermark from a watermarked image, the procedure presented in Fig. 5.5 is applied. Here $\hat{\mathbf{X}}$ denotes the watermarked image, whose content may be altered by natural noise or artifical modification, and $\mathbf{S}$ is the key used in the embedding procedure for selecting the necessary DCT bands. The steps followed illustrate how this procedure functions.

Step 1: Decompose $\hat{\mathbf{X}}$ into T non-overlapping blocks $\{\hat{\mathbf{x}}_1, \hat{\mathbf{x}}_2, ..., \hat{\mathbf{x}}_T\}$ with size 8×8 pixels.

Step 2: Perform the 8×8 block DCT (see Eq. (5.5)) on $\{\hat{\mathbf{x}}_1, \hat{\mathbf{x}}_2, ..., \hat{\mathbf{x}}_T\}$ one by one, and get the DCT coefficients $\{\hat{\mathbf{y}}_1, \hat{\mathbf{y}}_2, ..., \hat{\mathbf{y}}_T\}$ respectively.

Step 3: Generate the reference table $\hat{\mathbf{R}}$ from $\{\hat{\mathbf{y}}_1, \hat{\mathbf{y}}_2, ..., \hat{\mathbf{y}}_T\}$. With this table, we are able to determine what watermark bits are carried by the selected DCT bands.

Step 4: Determine which bands in $\hat{\mathbf{y}}_i$, $1 \leq i \leq T$, should be selected for extraction. Here the secret key $\mathbf{S}$ used in the embedding procedure should be presented for reference.

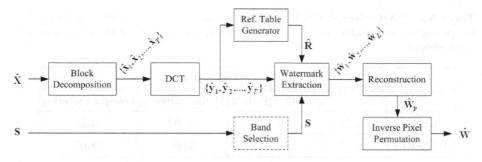

Fig. 5.5. Block diagram of the extracting procedure

Step 5: Extract the L hidden watermark bits $\{\hat{w}_i[1], \hat{w}_i[2], ..., \hat{w}_i[L]\}$ from the selected DCT bands. These extracted bits are collected as $\hat{\mathbf{w}}_i$.

Step 6: Repeat Steps 4 and 5 until all the DCT blocks have been handled. We have $\{\hat{\mathbf{w}}_1, \hat{\mathbf{w}}_2, ..., \hat{\mathbf{w}}_T\}$ afterwards.

Step 7: Piece together $\{\hat{\mathbf{w}}_1, \hat{\mathbf{w}}_2, ..., \hat{\mathbf{w}}_T\}$ to reconstruct a watermark image $\hat{\mathbf{W}}_P$.

Step 8: Apply the inverse pixel permutation procedure, see Eq. (5.8), to recover the original version of watermark $\hat{\mathbf{W}}$ from the permuted version $\hat{\mathbf{W}}_P$.

In the above procedure, details of most steps are same as those illustrated in the embedding procedure, thus they are omitted here. Please refer to them if necessary. In the extracting step (Step 5), to determine what watermark bit a selected DCT coefficient $\hat{y}_i[s[j]]$, where $1 \le i < T$ and $1 \le j < L$, is carrying, Eq. (5.14) is employed:

$$\hat{w}_i[j] = \begin{cases} 1, \text{ if } \hat{y}_i[s[j]] \cdot \hat{R}[s[j]] \ge \hat{y}_i[0]; \\ 0, \text{ otherwise.} \end{cases} \tag{5.14}$$

5.2.3 Simulation Results

To demonstrate the performance of the mentioned watermarking scheme, the well-known test images LENA (Fig. 1.2) and BABOON (Fig. 1.4) were used as the cover images. The image of ROSE (Fig. 5.3) was used as the watermark. The size of either cover image is 512×512 pixels in gray-level. The size of the watermark image is 128×128 pixels in bi-level. The key used for selecting DCT bands was generated randomly by following the definition, where the bands specified in one DCT block may be different from the others.

We decomposed both cover images into $T = \frac{512 \times 512}{8 \times 8} = 4096$ non-overlapping blocks. The watermark was then segmented into 4096 sections and each section contains $L = \frac{128 \times 128}{4096} = 4$ bits. Then, the steps mentioned in the embedding procedure were carried out with the above materials and settings.

To test the robustness of this DCT-based watermarking system, the JPEG compression, median filter and low-pass filter were employed to attack the watermarked images. The peak-signal-to-noise ratio (PSNR, see Eq. (2.4)) and the

Table 5.1. PSNR values and BCR values under the considered attacks when no training procedure is introduced and the images of LENA and BABOON are used as the cover images.

Cover image	PSNR (dB)	BCR (%)		
		JPEG, QF=60%	Median filtering	Low-pass filtering
LENA	30.33	98.14	60.97	57.63
BABOON	25.67	94.84	60.08	58.07

bit correct rate (BCR, see Eq. (2.7)) were applied to evaluate the performance in imperceptibility and robustness respectively.

Table 5.1 summarises the test results and shows the performance. Figures 5.6 and 5.7 show the watermarked result and the extracted results under the considered attacks respectively, while the image of LENA is used as the cover image. Figures 5.8 and 5.9 show the watermarked result and the extracted results under the considered attacks respectively, while the image of BABOON is used as the cover image.

5.2.4 Discussion

By comparing the watermarked results, which are Figs. 5.6 and 5.8, with the original cover images, which are Figs. 1.2 and 1.4, readers may have observed that some areas in the watermarked images have noticeable distortion. And, from the extratced results, we can see that the robustness under the attacks of median filtering and low-pass filtering is not good. This may also be caused by the same reason mentioned above.

For the first issue, the reason is the DCT bands selected in those areas for embedding are belonging to the lower-frequency part. As the energies of most natural images are concentrated in lower frequency bands, and human visual system is much more sensitive to lower frequencies, alteration in low-frequency bands thus is much eaiser to be noticed. However, if we modify high-frequency bands to carry watermark bits, the visual quality of the watermarked images may be good. But, since some image processing methods like low-pass filter will eliminate the higher frequencies of one image, the watermark bits carried by those high-frequency bands therefore will be filtered too. In other words, it means the robustness under this kind of attacks may not be good. And this is the reason for the second issue observed. In the literature, to serve as a trade-off for watermark embedding in the transform domain, most researchers suggest embedding the watermarks into the middle-frequency bands [37].

In conclusion, selecting the bands for embedding plays an important role in this DCT-based watermarking scheme. Thus, in next section we will focus on this topic and introduce a possible solution.

Fig. 5.6. The watermarked result when the image of LENA is used as the cover image.

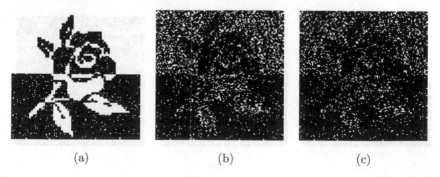

(a)   (b)   (c)

Fig. 5.7. The watermarks extracted from the watermarked LENA image attacked by (a) JPEG compression with a 60% quality factor, (b) median filtering, and (c) low-pass filtering

Fig. 5.8. The watermarked result when the image of BABOON is used as the cover image

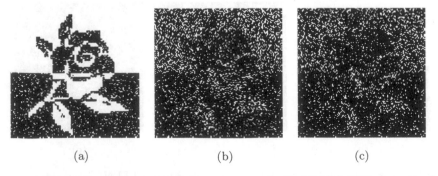

(a)	(b)	(c)

Fig. 5.9. The watermarks extracted from the watermarked BABOON image attacked by (a) JPEG compression with a 60% quality factor, (b) median filtering, and (c) low-pass filtering

5.3   Genetic Band Selection (GBS)

As mentioned previously in the DCT-based watermarking scheme, some DCT bands have to be selected for watermarking. This implies if the DCT bands can be selected well, the output result can have better visual quality or even stronger robustness. Thus, the ideas of designing a training procedure to select suitable DCT bands for better performance is formed. Here the training procedure, named *genetic band selection* (GBS), adapted from [77] is presented and introduced in this section. In this training procedure, an object function is defined and the genetic algorithm (GA) is employed to optimize it. The trained result is regarded as a secret key and is used in the DCT-based watermarking scheme introduced in Sect. 5.2.

5.3.1   Training Procedure

The definitions of the symbols used in Sect. 5.2 are now repeated. Let $\mathbf{X}$ be a cover image, $\mathbf{W}$ be a watermark to be embedded, and $\mathbf{S}$ be a user key defining which DCT bands should be selected. The cover image is first decomposed into T non-overlapping blocks $\{\mathbf{x}_1, \mathbf{x}_2, ..., \mathbf{x}_T\}$ of size 8×8 pixels, and the 8×8 block DCT is then applied on them respectively to obtain the DCT coefficients $\{\mathbf{y}_1, \mathbf{y}_2, ..., \mathbf{y}_T\}$. As the goal of the training procedure is to select a better set of DCT bands for embedding, the user key $\mathbf{S}$ therefore is regarded as a GA individual. Here we review the definition of $\mathbf{S}$ again.

Let $\mathbf{S}$ be a set containing T sequences $\{\mathbf{s}_1, \mathbf{s}_2, ..., \mathbf{s}_T\}$, where for any $i : 1 \leq i \leq T$, $\mathbf{s}_i$ contains L non-repeated integers $\{s_i[1], s_i[2], ..., s_i[L]\}$. That is, for any integers j and l ($1 \leq j \leq L$ and $1 \leq l \leq L$), $1 \leq s_i[j] \leq 63$ and $s_i[j] \neq s_i[l]$ if $j \neq l$. The L elements of $\mathbf{s}_i$ define which L DCT bands should be selected from $\mathbf{y}_i$. That is, $s_i[1]$ indicates the position of the first band to be selected, $s_i[2]$ indicates the position of the second band to be selected, and so on. As shown in the below figure, in an individual, each sequence is regarded as a chromosome and the integers within are regarded as its genes.

The procedures presented in the following sub-sections are then used as the training procedure.

A. General Training Procedure

Step 1: According to the definition of $\mathbf{S}$, generate N keys $\{\mathbf{S}_1, \mathbf{S}_2, ..., \mathbf{S}_N\}$ randomly as the initial GA individuals.

Step 2: Embed the watermark $\mathbf{W}$ into the cover image using these keys one by one. We thus have N watermarked images $\{\mathbf{X}'_1, \mathbf{X}'_2, ..., \mathbf{X}'_N\}$ then.

Step 3: Apply the considered attacks to $\{\mathbf{X}'_1, \mathbf{X}'_2, ..., \mathbf{X}'_N\}$ respectively. Here we use $\{\hat{\mathbf{X}}_1, \hat{\mathbf{X}}_2, ..., \hat{\mathbf{X}}_N\}$ to denote the attacked versions.

Step 4: Extract the hidden watermarks from $\{\hat{\mathbf{X}}_1, \hat{\mathbf{X}}_2, ..., \hat{\mathbf{X}}_N\}$ respectively. The extracted results are denoted by $\{\hat{\mathbf{W}}_1, \hat{\mathbf{W}}_2, ..., \hat{\mathbf{W}}_N\}$.

Step 5: Evaluate the performance of each individual according to its imperceptibility and robustness results.

Fig. 5.10. GA individuals

Step 6: Select the individuals with better performance.
Step 7: Stop the training procedure if the considered GA iteration t is met.
Step 8: Execute the mutation and crossover procedures to generate new individuals for next generation.
Step 9: Go to Step 2.

In the above procedure, the fitness function defined in Sect. 4.3.2.B may be borrowed and used in Step 5. That is:

$$f_i = f_{\mathrm{I}}(\mathbf{X}, \mathbf{X}'_i) + \lambda \times f_{\mathrm{R}}(\mathbf{W}_i, \mathbf{W}'_i), \; i = 1, 2, ..., N. \tag{5.15}$$

Here f_i denotes the performance of the i-th individual $\mathbf{S}_i$, f_{I} is the imperceptibility evaluating function (e.g. PSNR), f_{R} is the robustness evaluating function (e.g. BCR), and λ is a parameter used for controlling the balance between f_{I} and f_{R}. The fitness scores of the individuals are calculated one by one to measure their performance.

In Step 6, according to these scores, the probabilities for being selected as parents in the crossover step are assigned. Usually, the higher fitness score the individual has, the higher probability of becoming a parent the individual is assigned. According to the probabilities, the crossover process selects two parents to produce two children for next generation. After generating N new individuals, Step 8 is performed to mutate the genes of the new individuals according to a given mutation rate ρ_{m}. That is, for the j-th ($1 \le j \le N$) individual $\mathbf{S}_j$ and its i-th ($1 \le i \le L$) gene $s_j[i]$, if the probability generated by the random-number generator is less than ρ_{m}, then $s_j[i]$ is mutated as s', where $1 \le s' \le 63$ and for any integer $l : 1 \le l \le L$, $s' \ne s_j[l]$.

The above steps are repeated until the iteration t is satisfied. Finally, the individual with the best fitness score in the final generation is the output trained result.

B. Modified Training Procedure

The above training procedure is a whole-image-based procedure, which may be not so practical or effective in implementing it. Thus, a block-based training procedure modified from it is presented here.

Step 1: Decompose $\mathbf{X}$ into T non-overlapping blocks $\{\mathbf{x}_1, \mathbf{x}_2, ..., \mathbf{x}_T\}$ of size 8×8 pixels.

Step 2: Perform the 8×8 block DCT on $\{\mathbf{x}_1, \mathbf{x}_2, ..., \mathbf{x}_T\}$ respectively. We then have the transformed results $\{\mathbf{y}_1, \mathbf{y}_2, ..., \mathbf{y}_T\}$.

Step 3: Disperse the spatial relationships of $\mathbf{W}$ and decompose the permuted result into T non-overlapping blocks $\{\mathbf{w}_1, \mathbf{w}_2, ..., \mathbf{w}_T\}$. Here the size of either block is L.

Step 4: According to the definition of $\mathbf{S}$, generate N keys $\{\mathbf{S}_1, \mathbf{S}_2, ..., \mathbf{S}_N\}$ randomly as the initial individuals.

Step 5: Embed $\{\mathbf{w}_1, \mathbf{w}_2, ..., \mathbf{w}_T\}$ into $\{\mathbf{y}_1, \mathbf{y}_2, ..., \mathbf{y}_T\}$ respectively by referring to $\{\mathbf{s}_1^{(i)}, \mathbf{s}_2^{(i)}, ..., \mathbf{s}_T^{(i)}\}$, which is the content of the i-th $(1 \leq i \leq N)$ individual $\mathbf{S}_i$

Step 6: Apply the inverse DCT to the watermarked DCT blocks one by one. Here we use $\{\mathbf{x}_1'^{(i)}, \mathbf{x}_2'^{(i)}, ..., \mathbf{x}_T'^{(i)}\}$ to denote the reconstructed image blocks.

Step 7: Apply the considered attack to $\{\mathbf{x}_1'^{(i)}, \mathbf{x}_2'^{(i)}, ..., \mathbf{x}_T'^{(i)}\}$ respectively. We denote the attacked results as $\{\hat{\mathbf{x}}_1^{(i)}, \hat{\mathbf{x}}_2^{(i)}, ..., \hat{\mathbf{x}}_T^{(i)}\}$.

Step 8: Extract the hidden watermark bits from $\{\hat{\mathbf{x}}_1^{(i)}, \hat{\mathbf{x}}_2^{(i)}, ..., \hat{\mathbf{x}}_T^{(i)}\}$. We then have $\{\hat{\mathbf{w}}_1^{(i)}, \hat{\mathbf{w}}_2^{(i)}, ..., \hat{\mathbf{w}}_T^{(i)}\}$.

Step 9: Evaluate the performance of each chromosome using Eq. (5.15):

$$f_j = f_I(\mathbf{x}_j, \mathbf{x}_j'^{(i)}) + \lambda \times f_R(\mathbf{w}_j, \hat{\mathbf{w}}_j^{(i)}), \; j = 1, 2, ..., T. \quad (5.16)$$

We then have T scores to indicate the performance of $\{\mathbf{s}_1^{(i)}, \mathbf{s}_2^{(i)}, ..., \mathbf{s}_T^{(i)}\}$ respectively.

Step 10: Repeat Steps 5 to 9 to calculate the performance for next individual, until all the individuals have been evaluated.

Step 11: After the performance of all individuals are calculated, sort the scores and select the individuals with better performance.

Step 12: Stop the training procedure if the considered iteration t is met.

Step 13: Execute the mutation and crossover procedures to generate new N individuals for next generation.

Step 14: Go to Step 5.

In this modified procedure, since we skip the step of piecing together all image blocks to form a reconstructed image, the time used for training therefore can be reduced. But please note, the real performance with the trained result may have a little difference from the best performance of the final iteration. This is because while applying some image processing procedures to attack a block, the boundary pixels or the neighbor blocks are not considered. In summary, for a system where shorter training time is concerned, the modified procedure can be considered.

5.3.2 Simulation Results

To compare the watermarking results with and without the training procedure, the same materials and settings used in Sect. 5.2.3 were used in the simulation

Table 5.2. PSNR values and BCR values under the considered attacks when GBS is introduced and the images of LENA and BABOON are used as the cover images

Cover image	PSNR (dB)	BCR (%)		
		JPEG, QF=60%	Median filtering	Low-pass filtering
LENA	34.78	98.16	89.34	84.76
BABOON	28.10	94.42	90.65	89.72

again. Here in the training procedure, the block-based training procedure, as presented in Sect. 5.3.1.B, was employed and implemented. The settings we used were: the population size $N = 10$, the GA iteration $t = 100$, and the balance parameter $\lambda = 30$ and 25 while LENA and BABOON were used as the cover images respectively. At the beginning of the training procedure, all the bands recorded in the initial GA individuals were generated randomly. And, as the robustness under the attacks of low-pass filtering and median filtering was not good (see Table 5.1), we thus employed them in the attacking step and expected that the robustness could be improved.

Figure 5.11 shows the fitness scores of each iteration. Simulation results of the DCT-based watermarking scheme with the trained results, which is regarded as the secret keys, are summarised as Table 5.2. Figures 5.12 and 5.14 show the watermarked results while the images of LENA and BABOON were used as the cover images, respectively. Figures 5.7 and 5.9 show the extracted results while the considered attacks were employed to attack Figs. 5.12 and 5.14.

5.3.3   Comparison and Discussion

From the simulation results presented in Sects. 5.2.3 and 5.3.2, we can observe that the improvement in both the watermarked quality (PSNR values) and the robustness (BCR values) under certain attacks are improved, while the GBS training procedure introduced. The PSNR value increases from 30.33 dB to 34.78 dB while LENA is used as the cover image. And, the PSNR value increases from 25.67 dB to 28.10 dB while BABOON is used as the cover image. Also, we can see that the extarcted watermarks shown in Figs. 5.13 and 5.15 can be recognized clearly. With the aid of the GBS procedure, the performance actually improved largely.

As to the time required for training, we can see from Fig. 5.11 that the scores increase largely within the first 20 iterations. This means that even we only execute the training procedure for a short time, the trained result still can provide good enough performance. Other issues associated with the training procedure, such as key delivery, key reuse, attacks employed for tests, and the definiton of the fitness function used, sometimes may be important for a reason. We have already addressed these in Sect. 4.3.4 and thus they are skipped here. Readers may refer to them if needs.

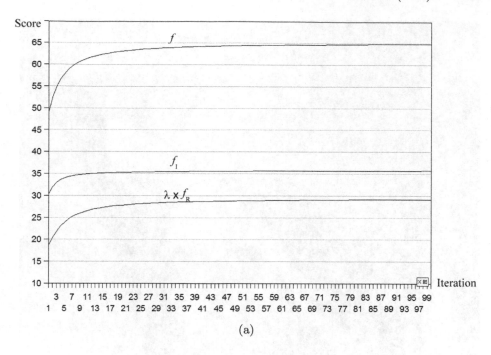

(a)

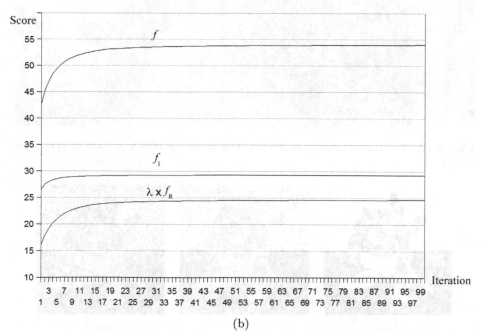

(b)

Fig. 5.11. Best scores recorded in each training iteration. (a) LENA is used as the cover image. (b) BABOON is used as the cover image.

Fig. 5.12. The watermarked result when the image of LENA is used as the cover image.

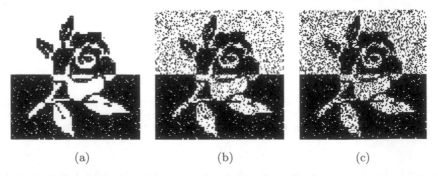

Fig. 5.13. The watermarks extracted from the attacked versions of Fig. 5.12, where the attacks are (a) JPEG compression with a 60% quality factor, (b) median filtering, and (c) low-pass filtering.

Fig. 5.14. The watermarked result when the image of BABOON is used as the cover image

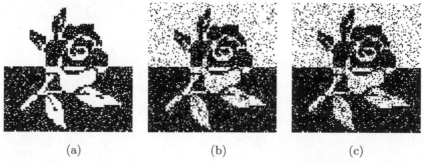

(a)   (b)   (c)

Fig. 5.15. The watermarks extracted from the attacked versions of Fig. 5.14, where the attacks are (a) JPEG compression with a 60% quality factor, (b) median filtering, and (c) low-pass filtering

5.4 Summary

A DCT-based watermarking scheme and a training procedure named genetic band selection (GBS) have been presented in this chapter. The watermarking scheme transforms the raw data of the cover image into frequency domain by applying DCT first. Then, it modifies the DCT bands selected by a user-specified key to carry the watermark bits. To improve the performance, the GBS training procedure is introduced. It employs genetic algorithms (GA) to find a better way to select a group of suitable DCT frequencies for watermarking. The trained result is therefore regarded as the secret key used in the DCT-based watermarking scheme. Simulation results presented have demonstrated that the visual quality improves. Further more, by taking the considered attacks into account, the trained result can also provide better resistance under those attacks considered, as the simulation results shown. In summary, with the aid of the GA-based training procedure, the performance of the original DCT-based watermarking scheme can be improved, just as what we expected.

Chapter 6
Vector Quantisation Based Watermarking Schemes and Codebook Partition

Watermarking techniques based on vector quantisation (VQ) have been getting more and more popular in recent years, since they enhance the traditional VQ systems by adding the ability of watermarking. In this chapter, our focus is on VQ-based watermarking schemes and the training procedure which improves the performance of the watermarking schemes introduced. We begin with the introduction of the general background in Sect. 6.1, then illustrate the traditional VQ coding method in Sect. 6.2. In Sect. 6.3 three watermarking schemes are presented. Experimental results, comparisons, and discussions are also included in this section. In Sect. 6.4, a training procedure named genetic codebook partition (GCP) is introduced. It employs the genetic algorithm (GA) to find a better way to split the codebook used, so that the trained result can be used in the mentioned VQ-based watermarking scheme to provide better watermarking results. Experimental results, comparisons, and discussions presented will hi-light its performance. Finally, Sect. 6.5 summarises this chapter.

6.1 Introduction

Compared with the watermarking techniques based on spatial domain and transform domain, the watermarking techniques based on quantisation domain only attracted attention in recent years. Among the quantisation techniques, vector quantisation (VQ) [24] is a well-known one, which has been applied on many areas successfully.

Focusing on the watermarking schemes based on VQ, Lu and Sun [52] introduced a method where they partitioned the original VQ codebook into a number of groups by referring to a secret key. They then modified the VQ indices according to the watermark bits and the groups partitioned. Their method requires the original cover image to be presented during extraction. To improve this shortage, Lu et al. proposed another method in [51]. However, this method requires expanding the codebook used. Jo and Kim [43] proposed a method to improve imperceptibility. They suggested partitioning the codebook used into

F.-H. Wang, J.-S. Pan, and L.C. Jain: Innovations in Dig. Watermark. Tech., SCI 232, pp. 83–107.
springerlink.com   © Springer-Verlag Berlin Heidelberg 2009

three groups according to a given threshold. They then used two of them in the watermark embedding procedure. In a similar way to [52], they modified the indices obtained to carry watermark bits.

Unlike the techniques which split the codebook used into groups for embedding, Huang et al. [39] [40] suggested hiding watermark bits in the output keys according to the VQ indices obtained. Their methods therefore can provide better imperceptibility. However, in their systems, the keys generated can only be used associating with the considered watermark and cover image. They cannot be reused. Wang et al. [97] proposed their method, which also splits the codebook used into groups, and utilities them to modify the VQ indices. So that these indices can carry watermark bits. They also proposed a training procedure which employs GAs to optimize the object function of how to split the codebook used better. By using the result of the training procedure in their watermarking scheme, the performance can be improved not only in imperceptibility, but also in robustness.

In Sect. 6.3, the methods proposed in [43] [52] [97] will be introduced. These schemes partition the codebook into a number of sub-codebooks (or groups) by reference to the user-key. Then, for each input vector of the cover image, a sub-codebook is selected according to the watermark bit to be embedded. The traditional VQ coding procedure is then done using the sub-codebook selected for the vector.

6.2 Vector Quantisation (VQ)

Vector quantisation [24] is an old coding method which has been applied successfully on image coding and speech coding. For image coding, the image to be coded is first decomposed into non-overlapping vectors (or, blocks). Then for each vector, a search procedure, named *nearest codeword search*, is performed to find a codeword with the minimum distortion from the codebook assigned. The index of the codeword obtained is afterwards transmitted to the receiver.

In the receiver, a table-lookup procedure using the same codebook as in the encoding procedure is carried out. According to the index received, the corresponding codeword from the codebook is output and regarded as the reconstructed vector. Block diagrams given in Fig. 6.1 illustrate the encoding and decoding procedures. In Fig. 6.1(a) $\mathbf{x}$ denotes a vector, $\mathbf{C} = \{\mathbf{c}_1, \mathbf{c}_2, ..., \mathbf{c}_L\}$ is a set (or, codebook) containing L predefined codewords, and I is and index, which is an integer and $1 \leq I \leq L$, of the nearest codeword obtained for $\mathbf{x}$.

In Fig. 6.1(b), $\mathbf{x}'$ denotes the reconstructed vector; $\{\mathbf{c}_1, \mathbf{c}_2, ..., \mathbf{c}_L\}$ and I have the same definitions as that in Fig 6.1(a). Details of the coding procedure are explained in the following sub-sections.

6.2.1 Codebook Design

For a VQ system, the codebook used plays an important role. A codebook is a set of codewords (or, vectors), and a codeword may contain a number of scalars

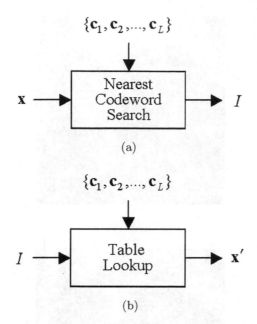

Fig. 6.1. Block diagrams of the VQ encoding and decoding procedures

$\{g_1, g_2, g_3, ..., g_n\}$ within. Generally speaking, the user of a VQ system can define his/her own codebook freely, and then uses this codebook to code the considered input data. But, this may result in varied coding performance in quality.

Figure 6.2 demonstrates two coded results using two different codebooks as examples. Here the image of BABOON (see 1.4) is decomposed into a number of non-overlapping blocks with size 4×4 pixels, and 256 blocks of them are selected randomly to form the first codebook. The same, the image of LENA (see 1.2) is decomposed into a number of non-overlapping blocks with size 4×4 pixels, and 256 blocks of them are selected randomly to form the second codebook. These two codebooks are used then to code Fig. 6.2(a). Figure 6.2(b) and Fig. 6.2(c) are the coded results respectively. Obviously, Fig. 6.2(c) has better visual quality than Fig. 6.2(b). This indicates that to yield better coded result in quality, a suitable codebook should be selected carefully.

Researchers have proposed many methods for generating codebooks. Among them, the Linde-Buzo-Gray (LBG) method [24] and the generalized Llyod algorithm (GLA) are the most well-known ones. However, due to this area about how to generate codebooks is not the main focus of this book, details of them thus are omitted. Readers who have the interests are encouraged to study them.

6.2.2 Encoding Procedure

In the encoding procedure, let $\mathbf{X}$ be an input image containing $W \times H$ pixels, and $\mathbf{C}$ be a codebook containing L codewords $\{\mathbf{c}_1, \mathbf{c}_2, ..., \mathbf{c}_L\}$ therein. Here for any

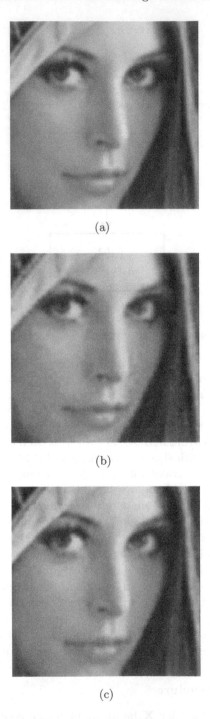

Fig. 6.2. VQ coded results using different codebooks. (a) Original image, (b) coded result using the 1st codebook, and (c) coded result using the 2nd codebook.

integer $i \in [1, L]$, $\mathbf{c}_i \in \mathbf{C}$ contains d pixels $\{c_1^i, c_2^i, ..., c_d^i\}$ therein. The following steps are used as the VQ encoding procedure:

Step 1: Decompose the given image $\mathbf{X}$ into T non-overlapping vectors $\{\mathbf{x}_1, \mathbf{x}_2, ..., \mathbf{x}_T\}$ of dimension d pixels. That is, $\mathbf{x}_i = \{x_1^i, x_2^i, ..., x_d^i\}$, where $1 \le i \le T$.

Step 2: Obtain a codeword with the minimum distortion from $\{\mathbf{c}_1, \mathbf{c}_2, ..., \mathbf{c}_L\}$ for each vector. Here we assume $\mathbf{c}_j = \{c_1^j, c_2^j, ..., c_d^j\}(1 \le j \le L)$ is the obtained codeword. The Euclidean distance is used to measure the distortion between two vectors. We have:

$$\mathrm{ED}(\mathbf{x}_i, \mathbf{c}_j) = \sum_{k=1}^{d}(x_k^i - c_k^j)^2. \qquad (6.1)$$

Usually, this step is called the nearest-codeword-search procedure.

Step 3: Collect the indices of all the codewords obtained as the index set $\mathbf{I}$. For example, if the nearest codeword to the i-th vector is $\mathbf{c}_8$, then its index, which is 8 (or 7 if counting is started from 0), is the i-th element of $\mathbf{I}$. The index set $\mathbf{I}$ is then transmitted to the decoder.

In the above procedure, instead of sending the obtained nearest codewords themselves to the decoder, the VQ system sends $\mathbf{I}$, which contains the indices of these codewords. The transmitted amount of data can then be reduced. Each vector has been compressed into $\log_2(L)/d$ of its original size, where $\log_2(L)$ is the length of an index. For example, for a given vector whose dimension is $d = 4 \times 4$ pixels, assume that the vector is in grey-level format (8 bits/pixel) and the codebook contains $L = 256$ codewords therein, then the VQ system uses $\log_2(L) = 8$ bits to express each vector (whose size is $4 \times 4 \times 8 = 128$ bits). Thus, the input vector is compressed into 8/128 which is 1/16 of its original size.

6.2.3   Decoding Procedure

To reconstruct an image from the VQ indices is not difficult. The steps below can be used. Here let $\mathbf{I} = \{I_1, I_2, ..., I_T\}$ be the received VQ index set and $\mathbf{C} = \{\mathbf{c}_1, \mathbf{c}_2, ..., \mathbf{c}_L\}$ be the codeword set used in the encoding procedure.

Step 1: Obtain the corresponding codeword from $\{\mathbf{c}_1, \mathbf{c}_2, ..., \mathbf{c}_L\}$ for each index. For example, for the i-th index I_i, where $1 \le i \le T$ and $1 \le I_i \le L$, the codeword $\mathbf{c}_{I_i}$ is selected and regarded as the output vector $\mathbf{x}'$. That is, $\mathbf{x}' = \mathbf{c}_{I_i}$.

This step is usually called the table-lookup procedure.

Step 2: Assemble all the output vectors $\{\mathbf{x}_1', \mathbf{x}_2', ..., \mathbf{x}_T'\}$ to form a reconstructed image $\mathbf{X}'$.

Examples of applying the VQ coding procedure to the image of LENA using different sizes of codebooks are given in Fig. 6.3.

(a) The original image of LENA

(b) $L = 256$ (PSNR=31.08 dB)

Fig. 6.3. Examples of applying the VQ coding procedure to the image of LENA when using difference sizes of codebooks

(c) $L = 128$ (PSNR=30.75 dB)

(d) $L = 64$ (PSNR=29.67 dB)

Fig. 6.3. (*continued*)

6.2.4 Summary

From the description presented in the decoding procedure, it shows that generating a decoded image from the indices received is easy, as only a table-lookup procedure is performed. In other words, the VQ decoding procedure is of low complexity and very effective. In most of the cases, the encoding work has to be executed once only and the decoding work has to be done for many times. This is another reason why the VQ system is a popular compression technique.

As to the shortcomings of the VQ system, the first is the coded quality. Comparing with the popular JPEG coding, the visual quality provided by the VQ coding is poorer. The second disadvantage is the delivery of the codebook used. As different users can use different codebooks to encode their images, it means in order to decode the VQ compressed results, the codebooks used in the encoding procedure have to be available. In other words, for a codebook containing L codewords where the size of each codeword is s bits, the encoder has to send $L \times s$ extra bits to the decoder. However, if both of the encoder and decoder use the same codebook to code images for a number of times, the codebook can be delivered only once. From another point of view, the VQ coding possesses better security, since without the correct codebook, the encoded data cannot be revealed correctly.

6.3 VQ-Based Watermarking Schemes

In this section, some VQ-based watermarking methods proposed in [52], [43] and [97] are illustrated briefly. To begin with, we define that $\mathbf{C}$ is a codebook containing L codewords, $\mathbf{X}$ is a cover image, and $\mathbf{W}$ is a binary bit stream to be embedded into $\mathbf{X}$. The information used for partitioning $\mathbf{C}$ into sub-codebooks is regarded as the secret key.

After decomposing $\mathbf{X}$ into T non-overlapping vectors $\{\mathbf{x}_1, \mathbf{x}_2, ..., \mathbf{x}_T\}$ of dimension d pixels, the normal VQ encoding procedure is performed to obtain the nearest codewords from $\mathbf{C}$ for all the vectors. The following sub-sections then illustrate how the related VQ-based methods utilize these codewords to carry the watermark bits.

6.3.1 Lu-Sun's Method [52]

For simplicity, the description below is limited to embedding only one watermark bit into each vector.

A. Original Method

In [52], the codebook $\mathbf{C}$ was partitioned into N groups $\{\mathbf{G}_1, \mathbf{G}_2, ..., \mathbf{G}_N\}$ where

(i) $\mathbf{C} = \bigcup_i^N \mathbf{G}_i$,
(ii) $\bigcap_i^N \mathbf{G}_i - \emptyset$,
(iii) $\mathbf{G}_i = \{\mathbf{c}_0^i, \mathbf{c}_1^i\}, 1 \leq i \leq N$, and
(iv) $N = L/2$.

The partitioned results are shown in Fig. 6.4.

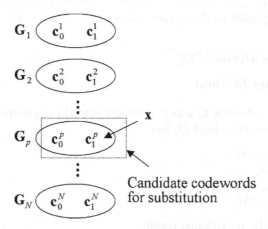

Fig. 6.4. Partitioning the codebook used into N groups by applying Lu-Sun's method, where $\mathbf{x}$ is a test vector

For a given vector $\mathbf{x}$, we assume that $c_g^p \in \mathbf{G}_p$, where $1 \leq p \leq N$ and $g \in \{0,1\}$, is the nearest codeword. To hide the corresponding watermark bit w in $\mathbf{x}$, the $(j+1)$-th codeword of $\mathbf{G}_p$ is given as the watermarked vector $\mathbf{x}'$:

$$j = (g + w) \text{ MOD } \| \mathbf{G}_p \|. \tag{6.2}$$

$$\mathbf{x}' = c_j^p. \tag{6.3}$$

Here $\| \mathbf{G}_p \|$ denotes the number of codewords contained in $\mathbf{G}_p$ (for embedding only one bit into each vector, $\| \mathbf{G}_p \| = 2$) and $0 \leq j < \| \mathbf{G}_p \|$. In Fig. 6.4, c_1^p is the nearest codeword to $\mathbf{x}$ (where $g = 1$). If $w = 0$, then c_1^p is used. If $w = 1$, c_0^p is used.

After all the watermark bits have been embedded into the corresponding vectors, the output vectors are pieced together to form the watermarked image $\mathbf{X}'$.

In addition, due to the embedding strategy used, Lu and Sun's method requires the original cover image to be presented during extraction, or otherwise the hidden information cannot be obtained.

B. Modified Method

To improve the disadvantage of requiring the original cover image during extraction, Eqs. (6.2) and (6.3), which are the embedding strategy, are modified. They then become Eqs. (6.4) and (6.5) respectively:

$$j = w. \tag{6.4}$$

$$\mathbf{x}' = c_w^p. \tag{6.5}$$

That is, after obtaining the nearest codeword from $\mathbf{C}$ for $\mathbf{x}$, which means p can be determined, and according to the bit w for embedding, the codeword $\mathbf{c}_w^p$ is selected and regarded as the output watermarked vector $\mathbf{x}'$.

6.3.2   Jo-Kim's Method [43]

A. Watermarking Method

In [43], the given codebook $\mathbf{C}$ was partitioned into three groups $\{\mathbf{G}_{-1}, \mathbf{G}_0, \mathbf{G}_1\}$ according to a given threshold D. Here

(i) $\mathbf{C} = \mathbf{G}_{-1} \cup \mathbf{G}_0 \cup \mathbf{G}_1$,
(ii) $\mathbf{G}_{-1} \cap \mathbf{G}_0 \cap \mathbf{G}_1 = \emptyset$,
(iii) $\mathbf{G}_i = \{\mathbf{c}_1^i, \mathbf{c}_2^i, ..., \mathbf{c}_M^i\}, i \in \{0, 1\}$, and
(iv) $L = ||\mathbf{G}_{-1}|| + 2M$.

Figure 6.5 shows the partitioned results.

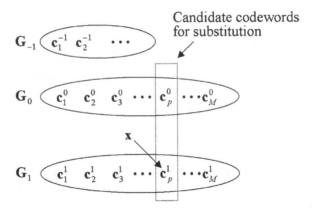

Fig. 6.5. Partitioning the codebook used into three groups by applying Jo-Kim's method, where $\mathbf{x}$ is a test vector.

For a given vector $\mathbf{x}$, we assume that $\mathbf{c}$ is the nearest codeword and w is the watermark bit for embedding. If $\mathbf{c} \in \mathbf{G}_{-1}$, then no watermark bit is embedded into $\mathbf{x}$. If $\mathbf{c} = \mathbf{c}_p^g \in \mathbf{G}_g$, where $g \in \{0, 1\}$ and $1 \le p \le M$, then the p-th codeword of $\mathbf{G}_w$ is output as the watermarked vector:

$$\mathbf{x}' = \mathbf{c}_p^w . \tag{6.6}$$

For instance, in Fig. 6.5, $\mathbf{c}_p^1$ is the nearest codeword to $\mathbf{x}$. To embed a bit-0 into $\mathbf{x}$, $\mathbf{c}_p^0$ is used; otherwise, $\mathbf{c}_p^1$ is used.

The above embedding process is repeated until all the watermark bits have been embedded into the input vectors. Afterwards, a watermarked image $\mathbf{X}'$ can be formed by assembling all the output vectors.

In addition, when assigning more codewords to $\mathbf{G}_{-1}$, which means a smaller threshold of D is used, the method provides better imperceptibility but lower embedding capacity.

B. Comparison with Lu-Sun's Methods

From Sects. 6.3.1 and 6.3.2, it can be seen that, if

(i) only one bit is embedded within each vector in [52] (that is, $\| \mathbf{G}_i \| = 2$ where $1 \le i \le N$), and

(ii) D is large enough in [43] (that is, $\| \mathbf{G}_{-1} \| = 0$ and $M = N$),

then Lu-Sun's method and Jo-Kim's method will have similar performances.

For example, we may collect $\{\mathbf{c}_0^1, \mathbf{c}_0^2, ..., \mathbf{c}_0^N\}$, the first codewords of all groups in Fig. 6.4, as the set $\mathbf{S}_0$, and collect $\{\mathbf{c}_1^1, \mathbf{c}_1^2, ..., \mathbf{c}_1^N\}$, the second codewords of all groups, as the set $\mathbf{S}_1$, as shown in Fig. 6.6. Comparing Fig. 6.5 with Fig. 6.6 and Eq. (6.5) with Eq. (6.6), it can be seen that both methods use the same strategy to utilize the vectors obtained for carrying watermark bits.

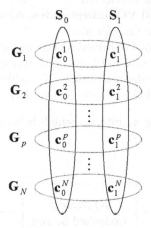

Fig. 6.6. Collecting the first codewords of all groups to form another group $\mathbf{S}_0$ and collecting the second codewords of all groups to form $\mathbf{S}_1$.

We can now say that [52] and [43] possess similar features and performances, but [43] provides an extra parameter D to control the balance between imperceptibility and embedding capacity.

6.3.3   Wang et al.'s Method [97]

A. Codebook Partition

For a given codebook $\mathbf{C}$ which contains L codewords $\{\mathbf{c}_1, \mathbf{c}_2, ..., \mathbf{c}_L\}$, a user-key $\mathbf{K} = \{k_1, k_2, ..., k_L | \forall k_i \in \{0,1\}, 1 \le i \le L\}$ is used to split it into two sub-codebooks $\mathbf{G}_0$ and $\mathbf{G}_1$. That is, the i-th ($1 \le i \le L$) codeword $\mathbf{c}_i$ is assigned to $\mathbf{G}_{k_i}$. The relationships among $\mathbf{C}$, $\mathbf{G}_0$, and $\mathbf{G}_1$ are

(i) $\mathbf{C} = \mathbf{G}_0 \cup \mathbf{G}_1$,
(ii) $\mathbf{G}_0 \cap \mathbf{G}_1 = \emptyset$.

Similar to [52] and [43], we use the codewords in $\mathbf{G}_0$ and $\mathbf{G}_1$ to hide a bit-0 and a bit-1 respectively.

B. Embedding Procedure

The steps for enabling the normal VQ system to have the watermarking ability are illustrated. They are:

Step 1: Partition the original codebook $\mathbf{C}$ into two sub-codebooks $\mathbf{G}_0$ and $\mathbf{G}_1$ by applying the method mentioned in Sect. 6.3.3A with the user-key $\mathbf{K}$.

Step 2: Divide the cover image $\mathbf{X}$ into T non-overlapping vectors $\{\mathbf{x}_1, \mathbf{x}_2, ..., \mathbf{x}_T\}$ of size d pixels.

Step 3: For the i-th $(1 \leq i \leq T)$ input vector $\mathbf{x}_i$ and the watermark bit w_i to be embedded into it, set $\mathbf{G}_{w_i}$ as the used codebook for the VQ system (i.e., if $w_i = 0$ then $\mathbf{G}_{w_i} = \mathbf{G}_0$ is selected).

Step 4: Execute the normal VQ nearest-codeword-search procedure to obtain a nearest codeword from $\mathbf{G}_{w_i}$ for $\mathbf{x}_i$.

Step 5: Output the codeword obtained as the watermarked vector $\mathbf{x}_i'$.

Step 6: Repeat Step 3 to Step 5 until all the watermark bits have been handled.

Step 7: Reconstruct a VQ decoded image $\mathbf{X}'$, which is also the watermarked image, by assembling all the output vectors $\{\mathbf{x}_1', \mathbf{x}_2', ..., \mathbf{x}_T'\}$.

The process of embedding w_i into $\mathbf{x}_i$, which is Step 3 to Step 5, is illustrated in Fig. 6.7.

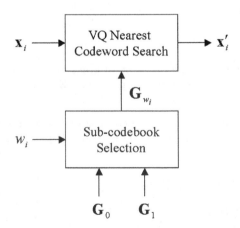

Fig. 6.7. The embedding procedure of Method [97]

C. Extraction Procedure

To extract the hidden watermark bits from a watermarked image received, the steps given below are applied.

Step 1: Decompose the received watermarked image $\hat{\mathbf{X}}$ into T non-overlapping vectors $\{\hat{\mathbf{x}}_1, \hat{\mathbf{x}}_2, ..., \hat{\mathbf{x}}_T\}$ of dimension d pixels.

Step 2: For an input vector $\hat{\mathbf{x}}_i$, where $1 \leq i \leq T$, execute the normal VQ nearest codeword search to obtain a nearest codeword from the original codebook **C**. Here we assume that $\mathbf{c}_j$, where $1 \leq j \leq L$, is the codeword obtained.

Step 3: Determine the watermark bit hidden in $\mathbf{c}_j$ according to its index j and the user-key **K**:

$$w'_i = k_j, \qquad (6.7)$$

where w'_i is the extracted watermark bit and k_j is the j-th element of **K**.

Step 4: Repeat Step 2 and Step 3 until all the hidden watermark bits $\{w'_1, w'_2, ..., w'_T\}$ have been extracted.

Step 5: Assemble $\{w'_1, w'_2, ..., w'_T\}$ to form the recovered watermark **W'**.

The process of extracting w'_i from $\hat{\mathbf{x}}_i$, which is Step 2 to Step 3, is illustrated in Fig. 6.8.

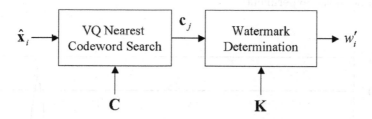

Fig. 6.8. The extraction procedure of Method [97]

In the above extraction procedure, splitting the original codebook into two sub-codebooks is unnecessary. The key **K** here is referred merely to indicate which codeword belongs to which sub-codebook. In other words, without the key which was used in the embedding procedure, no useful information can be extracted from the watermarked image.

6.3.4   Simulation Results and Comparison

In the simulation, the image of LENA (Fig. 6.3(a)) with size 512×512 pixels in grey-level was used as the cover image. The image of ROSE (Fig. 1.1(a)) with size 128×128 pixels in bi-level was used as the watermark. A codebook containing $L = 256$ codewords was obtained from the image of LENA by applying the LBG algorithm [24] with a threshold of 0.0001. The cover image was decomposed into

16384 non-overlapping blocks ($T = 16384$) of size 4×4 pixels ($d = 16$). To evaluate the performance of the introduced systems, the PSNR (Eq. (2.4)) and the BCR (Eq. (2.7)) were employed to examine the imperceptibility and the robustness respectively. In addition, without embedding any extra bit into the cover image, the PSNR value between the cover image and the VQ coded image is 31.80 dB.

A. Imperceptibility Test

To determine an objective performance of the introduced watermarking schemes, 10000 experiments were done. In each of the experiments, a codebook-partition key was generated randomly to partition the original codebook into two sub-codebooks, where $\| \mathbf{G}_0 \| = \| \mathbf{G}_1 \| = 128$. As a comparison, Jo-Kim's method was also run for 10000 times with the same test data and a threshold of $D = 100$. As mentioned previously, Lu-Sun's method and Jo-Kim's method have similar performances, therefore here we only compare Jo-Kim's method with Wang et al.'s method.

Figure 6.9 displays the PSNR values of the 10000 experiments. Table 6.1 lists the performance of both methods, where AEB denotes the average embedded bits, APSNR denotes the average PSNR value, and BPSNR denotes the best PSNR value of the experiments.

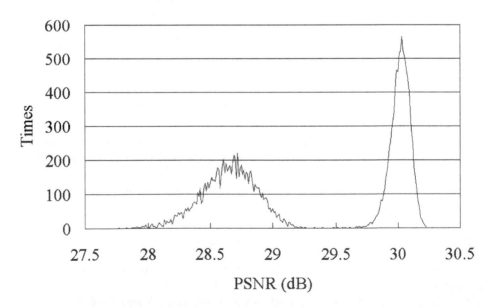

Fig. 6.9. Experimental results of 10000 experiments using Jo-Kim's scheme (left) and Wang et al.'s scheme (right)

Table 6.1. Imperceptibility test results of the two methods

Item	Jo-Kim [43]	Wang et al. [97]
BCR	100%	100%
AEB	16112 bits	16384 bits
APSNR	28.65 dB	30.02 dB
BPSNR	29.39 dB	30.23 dB

B. Robustness Test

To test the robustness of the introduced watermarking schemes, some common image-processing methods listed in Table 6.2 were employed to attack the watermarked images. From the 10000 experiments, one experiment is selected randomly and its test results are listed here. For Jo and Kim's method, Table 6.3 lists the embedded results when different values of threshold D are used. Tables 6.4, 6.5, and 6.6 display the extracted results when the threshold D is 50, 100, and 200, respectively. For Wang et al.'s method, the PSNR value between the cover image and the watermarked image is 29.78 dB. Figure 6.10 displays the watermarks extracted from the attacked images and Table 6.2 lists the robustness test results.

Table 6.2. Robustness test results of the Wang et al.'s scheme

Attack	BCR (%)	Extracted bits
VQ	100.00	16384
JPEG, QF=40%	88.64	16384
JPEG, QF=60%	95.37	16384
JPEG, QF=80%	99.80	16384
Low-pass filtering	80.44	16384
Median filtering	91.99	16384
Cropping, 25%	86.13	16384
Shifting, downward 1 line	82.79	16384

Table 6.3. The embedded results of Jo and Kim's method, where $g \in \{0, 1\}$

D	$\| \mathbf{G}_{-1} \|$	$\| \mathbf{G}_g \|$	PSNR (dB)	Embedded bits
50	110	73	30.38	14365
100	18	119	28.96	16223
200	0	128	28.28	16384

Table 6.4. Robustness test results of Jo and Kim's method when $D = 50$

Attack	BCR (%)	Extracted bits
VQ compression	100.00	14365
JPEG, QF=40%	50.76	14429
JPEG, QF=60%	51.17	14375
JPEG, QF=80%	51.41	14367
Low-pass filtering	51.51	14802
Median filtering	55.40	14602
Cropping, 25%	74.08	15245
Shifting, downward 1 line	50.46	14437

Table 6.5. Robustness test results of Jo and Kim's method when $D = 100$

Attack	BCR (%)	Extracted bits
VQ compression	100.00	16223
JPEG, QF=40%	65.31	16226
JPEG, QF=60%	94.64	16223
JPEG, QF=80%	99.64	16223
Low-pass filtering	53.44	16287
Median filtering	55.69	16258
Cropping, 25%	74.16	16281
Shifting, downward 1 line	54.18	16249

Table 6.6. Robustness test results of Jo and Kim's method when $D = 200$

Attack	BCR (%)	Extracted bits
VQ compression	100.00	16384
JPEG, QF=40%	86.68	16384
JPEG, QF=60%	94.82	16384
JPEG, QF=80%	99.65	16384
Low-pass filtering	73.11	16384
Median filtering	85.26	16384
Cropping, 25%	88.87	16384
Shifting, downward 1 line	76.29	16384

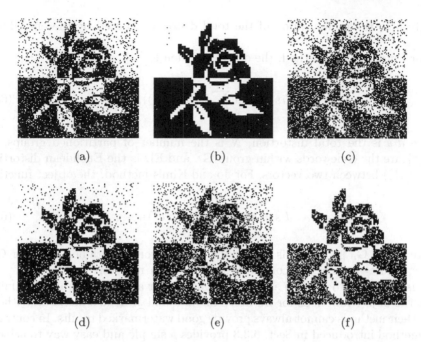

Fig. 6.10. The watermarks extracted from the attacked images: (a) JPEG compression with a QF=40%, (b) JPEG compression with a QF=80%, (c) low-pass filtering with a window size=3, (d) median filtering a with window size=3, (e) shifting 1 line downward, and (f) cropping 25% in the lower-left quarter.

6.3.5   Discussions

Generally speaking, Wang et al.'s watermarking method has better performance in encoding time, image quality, and robustness compared with the related VQ-based watermarking methods. This section presents the explanations and some issues relating to our findings.

A. Encoding Time

In the embedding procedures of the related VQ-based watermarking schemes such as [39] [40] [43] [52], they search the whole codebook to obtain a nearest codeword for each input vector. By contrast, Wang et al.'s method achieves this job by merely using a sub-codebook. The watermarking scheme therefore possesses shorter encoding time than the other methods listed.

B. Codebook Partition Methods

As stated in Sects. 6.3.1, 6.3.2, and 6.3.3, the method used for partitioning code-books plays an important role. Using a poor partitioned result for the embedding procedure leads to a poor watermarked result. To ensure the partition has minimum effect on the embedding method, it was found that consideration must be given to the minimisation of the total distortion caused by the embedding process.

For Lu and Sun's method, the object function is:

$$dis = \text{Min}(\sum_i^N \text{ED}(\mathbf{c}_0^i, \mathbf{c}_1^i)), \tag{6.8}$$

where dis is the total distortion, N is the number of partitioned groups, $\mathbf{c}_0^i$ and $\mathbf{c}_1^i$ are the codewords within group $\mathbf{G}_i$, and ED is the Euclidean distortion (Eq. (2.1)) between two vectors. For Jo and Kim's method, the object function is:

$$dis = \text{Min}(\sum_i^M \text{ED}(\mathbf{c}_i^0, \mathbf{c}_i^1)), \tag{6.9}$$

where $M =\parallel \mathbf{G}_0 \parallel = \parallel \mathbf{G}_1 \parallel$ is the number of codewords contained in $\mathbf{G}_0$ or $\mathbf{G}_1$, and $\mathbf{c}_i^0$ and $\mathbf{c}_i^1$ are the i-th codeword of $\mathbf{G}_0$ and $\mathbf{G}_1$ respectively.

It is necessary to develop a suitable algorithm or a training strategy to minimize the object functions for Lu-Sun's method and Jo-Kim's method. Without this, their methods cannot always provide good watermarked results. In contrast, the method introduced in Sect. 6.3.3 provides a simple and easy way to achieve the partition job effectively. Experimental results of the 10000 experiments have verified this.

C. Embedding Methods

The results from Fig. 6.9 and Table 6.1 show that Wang et al.'s watermarking scheme provides better imperceptibility even using no training technique. The reason for this is clear: in that the embedding procedure, while searching for a codeword to substitute the original nearest codeword for hiding the watermark bit, Wang et al.'s method provides more candidate codewords for choosing, unlike Lu-Sun's method and Jo-Kim's method, which only provide a default codeword for substitution, irrespective whether this default codeword is close to the input vector or not.

As shown in Figs. 6.4 and 6.5, after obtaining the nearest codeword for an input vector $\mathbf{x}$, only the default codeword, which is $\mathbf{c}_0^p$ or $\mathbf{c}_1^p$ in Fig. 6.4 and $\mathbf{c}_p^0$ or $\mathbf{c}_p^1$ in Fig. 6.5, can be used for substitution. In contrast, all the codewords in the corresponding group are available candidates, as shown in Fig. 6.11. To obtain another codeword with a smaller distortion for substitution is more practical than either Lu-Sun's method and Jo-Kim's method.

D. Robustness of Using One Vector to Carry Multi-Bits

Generally speaking, the watermarking schemes which divide the input image into blocks for embedding or extraction have poorer robustness under some kinds of attacks such as cropping or rotation. An example is now given.

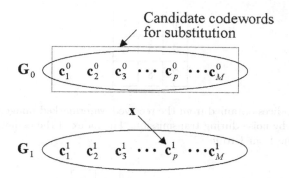

Fig. 6.11. The illustration of more candidate codewords are provided for substitution in Wang et al.'s method, where **x** is a test vector.

1	2	3		-	0	1
4	5	6		0	1	0
7	8	1		1	-	-

(a)	(b)

Fig. 6.12. The example of employing Jo and Kim's method to a given image. (a) The modified VQ indices. (b) The bits embedded in the image, where the symbol '-' denotes no watermark bit is embedded in the block.

For a given codebook $\mathbf{C}$, Jo and Kim's method is carried out to split it into three sub-codebooks $\{\mathbf{G}_{-1}, \mathbf{G}_0, \mathbf{G}_1\}$. For simplicity, we assume that $\mathbf{C} = \{c_1, c_2, ..., c_8\}$, $\mathbf{G}_{-1} = \{c_1, c_8\}$, $\mathbf{G}_0 = \{c_2, c_4, c_6\}$, and $\mathbf{G}_1 = \{c_3, c_5, c_7\}$. After employing Jo and Kim's method to a given image with the above partition information, the modified VQ indices and the embedded bits are shown in Fig. 6.12. As described in Sect. 6.3.2, if the nearest codeword for a given vector belongs to $\mathbf{G}_{-1}$, no watermark bit is embedded into the current vector. The generated watermarked image is afterwards transmitted to the receiver by some channel. We assume that some noise is added to the watermarked image during the transmission.

When the receiver has the watermarked image, Jo and Kim's method is applied. The indices obtained are shown in Fig. 6.13, where $1 \le y \le 8$. Here if $c_y \in \mathbf{G}_{-1}$ (e.g., $y = 1$ or $y = 8$), the bits embedded will be extracted correctly without distortion, such as Fig. 6.14(a). If $c_y \in \mathbf{G}_g$ (e.g., $y = 2$), where $g \in \{0, 1\}$, the bits extracted will be incorrect, such as Fig. 6.14(b).

From the above example, it is clear that the watermark bits are easy to shift when noise occurs. This problem is due to the use of $\mathbf{G}_{-1}$. In other words, due to using a single vector to carry vary number of watermark bits, as in Lu and Sun's method. It explains why the extracted results in Tables 6.4 to 6.6 are not good under some common attacks.

y	2	3
4	5	6
7	8	1

Fig. 6.13. The indices obtained from the received watermarked image, where the first vector is affected by noise during transmission. The index of the codeword obtained is therefore becoming from 1 to y.

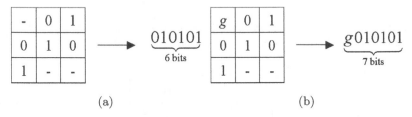

(a)   (b)

Fig. 6.14. The bits extracted from the watermarked image containing distortion when: (a) $c_y \in G_{-1}$ (e.g., $y = 1$) and (b) $c_y \in G_g$ (e.g., $y = 2$), where $g \in \{0, 1\}$.

E. User-Key Selection, Reuse, and Delivery

In a watermarking system, issues such as whether the users should have the full control to select or define the user-keys, whether the keys can be reused, or whether delivering the keys is important, are sometimes of concern but sometimes not. Usually these issues depend on the application where the watermarking system is applied. Among the above issues, the key delivery issue is beyond the focus of this chapter. It is therefore omitted and the focus is on the other two issues.

As mentioned in Sect. 6.3.5B, most related VQ-based watermarking schemes require a suitable technique to generate the used keys. For example, the paper [52] mentions a tabu search approach was used to obtain the codebook-partition key. A comparison regarding to the keys used is given in Table 6.7.

6.3.6 Summary

Three VQ-based watermarking schemes which partition the codebook used into a number of sub-codebooks, and then use these sub-codebooks for embedding and extraction have been introduced. These schemes are not difficult to implement and can enhance the traditional VQ system the ability of watermarking. Compared with the three watermarking schemes, Wang et al.'s scheme possesses the advantages of better imperceptibility, stronger robustness, shorter encoding time, and complete freedom for users to choose their own secret keys. Experimental results presented have shown the superiority of these approaches.

Table 6.7. The comparison of the keys used among some VQ-based watermarking methods

Scheme	Purpose of key(s)	Generating methods	Reusable or not
Lu-Sun [52]	Partition the used codebook	Training techniques	Reusable if associated with the used codebook.
Huang-Wang-Pan [39] [40]	Extract hidden watermarks	Generated from the cover image and the watermark	Not reusable. Associated with the cover image and watermark only.
Jo-Kim [43]	Partition the used codebook	Training techniques	Reusable if associated with the used codebook.
Wang et al. [97]	Partition the used codebook	Random-assigned or training techniques	Reusable.

6.4 Genetic Codebook Partition (GCP)

For VQ-based watermarking schemes such as [43] [52], as the investigation in [29], partitioning the original codebook into sub-codebooks or groups plays an important role. Therefore, in this section the method for partitioning the codebook used becomes the focal point. A genetic codebook partition (GCP) scheme is presented and described. It employs GAs to find a better way to split the given codebook. The imperceptibility of the VQ-based watermarking scheme can be improved using this procedure.

6.4.1 Training Scheme

The definitions of the symbols used in Sect. 6.3 are now repeated. Let $\mathbf{X}$ be a cover image which is a set of $M \times H$ pixels. $\mathbf{C}$ is a codebook which is a set of L codewords $\{\mathbf{c}_1, \mathbf{c}_2, ..., \mathbf{c}_L\}$. $\mathbf{K}$ is a user key which is a set of L bits $\{k_1, k_2, ..., k_L\}$. As described in Sect. 6.3.3, $\mathbf{K}$ is used to split $\mathbf{C}$ into two groups $\mathbf{G}_0$ and $\mathbf{G}_1$, where $\mathbf{C} = \mathbf{G}_0 \cup \mathbf{G}_1$ and $\mathbf{G}_0 \cap \mathbf{G}_1 = \emptyset$. That is, for a given codeword $\mathbf{c}_i$, where $1 \leq i \leq L$ and $\mathbf{c}_i \in \mathbf{C}$, it is assigned to group $\mathbf{G}_{k_i}$, where k_i is the i-th bit of $\mathbf{K}$.

A. Steps

The steps below are used in the GCP procedure.

Step 1: Generate S keys $\{\mathbf{K}_1, \mathbf{K}_2, ..., \mathbf{K}_S\}$ randomly as the initial GA individuals.
Step 2: Embed $\mathbf{W}$ into $\mathbf{X}$ using $\{\mathbf{K}_1, \mathbf{K}_2, ..., \mathbf{K}_S\}$, respectively.

Step 3: Evaluate the performance of each individual by calculating their fitness scores in the given fitness function.

Step 4: Select the individuals with better fitness scores and store this information.

Step 5: Output the best individual and terminate the training procedure if the required iteration criterion is met. Otherwise, keep executing the following steps.

Step 6: Execute the crossover procedure to generate S new individuals for next generation.

Step 7: Mutate the genes of the new individuals.

Step 8: Go to Step 2.

B. Details of the GCP Procedure

In Step 1, the goal of the GCP procedure is to find a better way to split the codebook used, the user key $\mathbf{K}$ therefore is regarded as a GA individual. We generate S keys randomly as the initial GA individuals. For instance, if $S = 3$ and $L = 6$, then S keys with length L bits such as $\mathbf{K}_1 = \{0, 1, 1, 0, 1, 0\}$, $\mathbf{K}_2 = \{0, 1, 0, 1, 0, 1\}$, and $\mathbf{K}_3 = \{1, 1, 0, 0, 1, 0\}$ can be used.

In the embedding step (Step 2), the VQ-based watermark embedding procedure introduced in Sect. 6.3.3 is employed. Each generated key is used to split the codebook into two groups so that $\mathbf{W}$ can be embedded into $\mathbf{X}$ according to the partitioned results. By repeating this embedding procedure for S times, where each time a different key is used, then S watermarked images $\{\mathbf{X}'_1, \mathbf{X}'_2, ..., \mathbf{X}'_S\}$ are generated.

The performance for each individual is then evaluated according to a given fitness function such as:

$$f_i = f_{\mathrm{I}}(\mathbf{X}, \mathbf{X}'_i), \; i = 1, 2, ..., S, \tag{6.10}$$

where f_i denotes the performance of $\mathbf{K}_i$ and f_{I} denotes the evaluating function for imperceptibility (e.g. PSNR). The reason for using this fitness function is explained in Sect. 6.4.3. By calculating the fitness function for each watermarked image, S scores $\{f_1, f_2, ..., f_S\}$ are obtained. These scores denote the respectively performances of $\{\mathbf{K}_1, \mathbf{K}_2, ..., \mathbf{K}_S\}$. The higher value the fitness score, the better performance of the individual. According to these scores, the probabilities for being selected as parents in the crossover step are assigned. Similarly to the GPS procedure, we give an individual a higher probability if it has a better performance. The crossover step finally generates S new individuals for next generation.

In Step 7, the genes of these new individuals are mutated according to the given mutation rate ρ_{m}. That is, if for the i-th gene k_i^j ($1 \le i \le L$) of a given individual $\mathbf{K}_j$ ($1 \le j \le S$), the probability generated by the random-number generator is less than ρ_{m}, k_i^j is then mutated as $1 - k_i^j$. This step means the corresponding codeword is removed from group $\mathbf{G}_{k_i^j}$ and assigned to another group $\mathbf{G}_{(1-k_i^j)}$.

The above steps are repeated until the considered iteration t is satisfied. Finally, the individual with the best fitness score in the final generation is the output trained result, as described in Step 5. The output result of the GCP procedure can be used in the mentioned VQ-based watermarking scheme to improve the imperceptibility of the watermarking scheme.

6.4.2 Simulation Results and Comparisons

To compare the performance of the VQ-based watermarking scheme with and without the GCP procedure, the data used in Sect. 6.3.4 were reused again. They are: the image of LENA (Fig. 1.2) was used as the cover image, the image of ROSE (Fig. 1.1(a)) was used as the watermark, and a codebook with 256 codewords ($L = 256$) was used. For the GCP procedure, the settings of $S = 10$, $t = 1000$, $\rho_s = 100\%$, $\rho_c = 50\%$, and $\rho_m = 0.1\%$ were used. To test the robustness, the same attack schemes were also applied.

Table 6.8 lists the imperceptibility test results with and without the GCP procedure. Table 6.9 lists the robustness of the test results with and without the GCP procedure. For demonstration, Fig. 6.15 displays the watermarks extracted from the attacked watermarked images when the GCP procedure is introduced.

Table 6.8. The imperceptibility test results (PSNR) without and with the GCP procedure

Method	PSNR (dB)
Without GCP	29.78
With GCP	30.42

Table 6.9. The robustness test results (BCR) without and with the GCP procedure

Attack	BCR (%)	
	Without GCP	With GCP
VQ	100.00	100.00
JPEG, QF=40%	88.64	84.76
JPEG, QF=60%	95.37	94.41
JPEG, QF=80%	99.80	99.78
Low-pass filtering	80.44	82.64
Median filtering	91.99	93.94
Cropping, 25%	86.13	86.82
Shifting, downward 1 line	82.79	82.89

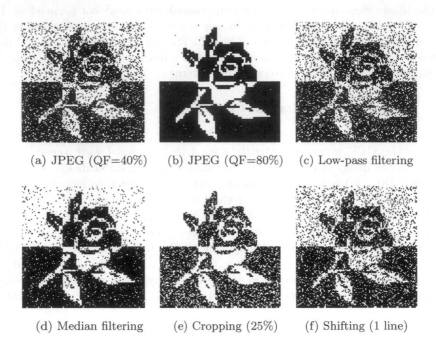

(a) JPEG (QF=40%)   (b) JPEG (QF=80%)   (c) Low-pass filtering

(d) Median filtering   (e) Cropping (25%)   (f) Shifting (1 line)

Fig. 6.15. The watermarks extracted from the attacked images when the trained result of the GCP procedure is used.

6.4.3   Discussion

In the VQ system, for two input vectors with a reasonable distortion between them, the VQ system will usually obtain the same codeword for them. For example, let $\mathbf{x}'$, $\hat{\mathbf{x}}$, and Search() denote the watermarked vector, the watermarked vector containing noise, and the VQ nearest-codeword-search procedure, respectively. Usually, if $\mathbf{x}'$ is similar to $\hat{\mathbf{x}}$, then Search($\mathbf{x}'$) = Search($\hat{\mathbf{x}}$). Therefore, we are more concerned how to obtain the codewords with the minimum distortion for the input vectors, which is imperceptibility, than the robustness. To simplify the GA training procedure, we only considered the PSNR as the fitness function in the proposed GCP procedure.

However, in some cases the robustness against some considered attacks sometimes is important. Therefore, the fitness functions, such as Eqs. (4.21) and (4.22) used in Sects. 4.3.2.B and 4.3.4.C, can be borrowed in order to take the effect of the possible attacks into account.

In conclusion, with the aid of the GCP procedure, the performance of the VQ-based watermarking scheme is improved. Experimental results support the superiority of this scheme.

6.5   Summary

In this chapter, three vector quantisation based watermarking schemes and a training procedure named genetic codebook partition (GCP) have been introduced. These watermarking schemes all split the codebook first and use the split sub-codebooks to hide the watermark bits. From the illustration readers should be able to know how this kind of watermarking systems embed watermark bits into the cover image by modifying the VQ indices obtained. Experimental results, comparisons, and discussions presented have demonstrated their performance and characteristics. Also, the introduced GCP procedure, which employs genetic algorithms to find a better way to partition the codebook into the considered number of sub-codebooks for watermarking, shows a way to optimize the performance of the schemes mentioned. Experimental results given have demonstrated this. In summary, from the illustration presented, readers now should have the basic ideas about how a classical VQ-based watermarking scheme works and what a training procedure can provide.

6.5 Summary

Chapter 7
Genetic Index Assignment

To increase the capacity of a watermarking system, compressing the original
watermark into a smaller size is one of the reasonable solutions. In this chapter,
a VQ-based coding procedure [100] [101] with an index-assignment procedure is
illustrated for general watermarking schemes to enlarge their capacity. It employs
the VQ operator to compress the gray watermark and adjusts the compressed
result according to an index table. Then, a genetic index assignment (GIA) [96]
procedure is introduced to find a better way to assign the indices in order to
have better imperceptibility.

7.1 VQ-Based Gray Watermark Coding Procedure

A coding procedure for gray watermarks proposed in [101] [100] is illustrated
in this section. It utilizes the VQ procedure to compress the gray watermark as
a binary one. This procedure is described as follow.

7.1.1 Gray Watermark Coding

Let $\mathbf{W}_G$ be a gray watermark and $\mathbf{C}$ be a codebook containing L codewords. After decomposing $\mathbf{W}_G$ into T non-overlapping vectors of size d pixels and applying
the VQ encoding procedure (Sect. 6.2.2) to them, the indices $\{I_1, I_2, ..., I_T | \forall I_i \in
[0, L), i \in [1, T]\}$ of the nearest codewords obtained for the corresponding vectors
are collected as the index set $\mathbf{I}$. Next, $\mathbf{I}$ is translated into a binary bit stream
$\mathbf{W}_B$ according to the value of L. For example, if $L = 16$ (which means the length
of each index is $\log_2(L) = 4$ bits) and $I_1 = 5$, then I_1 is translated as $\{0,1,0,1\}$
because $(5)_{10} = (0101)_2$. The bits $\{0,1,0,1\}$ then form the first four bits of $\mathbf{W}_B$.
The above process is repeated until all the indices in $\mathbf{I}$ have been translated.

The whole encoding procedure may be summarised as Eq. (7.1) and its inverse
procedure may be summarised as Eq. (7.2). An example of applying this coding
procedure to a given gray watermark is shown in Fig. 7.1, where a codebook
containing $L = 16$ codewords is used.

F.-H. Wang, J.-S. Pan, and L.C. Jain: Innovations in Dig. Watermark. Tech., SCI 232, pp. 109–117.
springerlink.com   © Springer-Verlag Berlin Heidelberg 2009

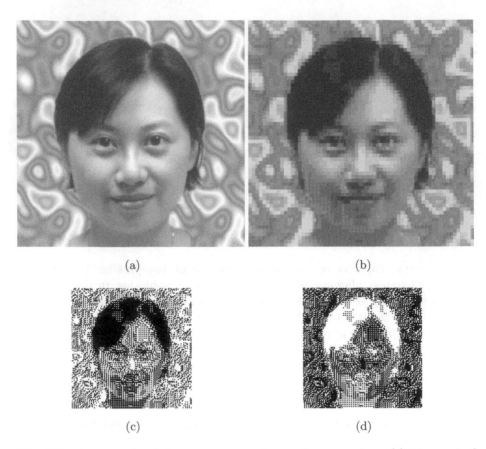

Fig. 7.1. An example of the gray watermark encoding procedure. (a) The original gray watermark ($256 \times 256 \times 8$ bits/pixel). (b) The VQ coded result ($256 \times 256 \times 8$ bits/pixel). (c) The binary bit stream, which is arranged as a binary image of size 128×128 pixels, obtained from the VQ indices of (b). (d) The binary bit stream, which is also arranged as a binary image of size 128×128 pixels, obtained from the indices of (b) when different indices are assigned to the codewords.

$$\mathbf{W}_B = \text{Encode}(\mathbf{W}_G, \mathbf{C}) \ . \tag{7.1}$$

$$\mathbf{W}_G = \text{Decode}(\mathbf{W}_B, \mathbf{C}) \ . \tag{7.2}$$

7.1.2 Index Assignment

In Sect. 7.1, a codebook $\mathbf{C} = \{\mathbf{c}_1, \mathbf{c}_2, ..., \mathbf{c}_L\}$ is used to encode the gray watermark $\mathbf{W}_G$. When no index assignment is introduced, the indices of $\{\mathbf{c}_1, \mathbf{c}_2, ..., \mathbf{c}_L\}$ are $\{0, 1, ..., L-1\}$ respectively. If we assign new indices to all the codewords as

$\{L-1, L-2, ..., 0\}$, for example, then the binary bit stream generated from $\mathbf{W}_G$ by applying Eq. (7.1) will differ from the original one.

Figure 7.2 illustrates assigning different indices to the codewords, where it is assumed that $L = 16$ and the nearest codeword obtained for a given vector of $\mathbf{W}_G$ is c_2. When no index assignment is introduced, the index of c_2, which is 1 (Fig. 7.2(a)), is translated as $\{0, 0, 0, 1\}$ using the method introduced in Sect. 7.1. After assigning new indices $\{L-1, L-2, ..., 0\}$ to the codewords (Fig. 7.2(b)), the index of c_2 now is 14. It is then translated as $\{1, 1, 1, 0\}$. Figure 7.1(d) shows the example of the gray watermark coded result when index assignment is introduced and indices $\{L-1, L-2, ..., 0\}$ are assigned.

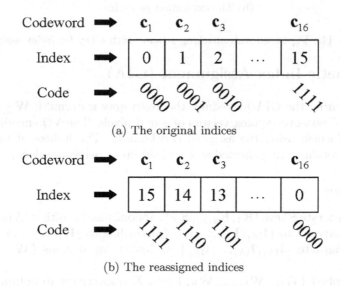

(a) The original indices

(b) The reassigned indices

Fig. 7.2. Coded results while assigning different indices to the codewords

7.1.3   Watermarking Scheme with Index Assignment

Based on the concepts of Sects. 7.1.1 and 7.1.2, a watermarking scheme with an index assignment procedure for gray watermarks can be designed as Fig. 7.3. Here $\mathbf{K}$ is an index-assignment key for assigning new indices to the codewords in $\mathbf{C}$. For instance, $\mathbf{K} = \{0, 1, ..., 15\}$ in Fig. 7.2(a) and $\mathbf{K} = \{15, 14, ..., 0\}$ in Fig. 7.2(b). The other symbols used in Fig. 7.3 will be explained in the following sections.

In this watermarking scheme, as it employs the VQ operator to compress the gray watermark, the capacity of the scheme is enlarged.

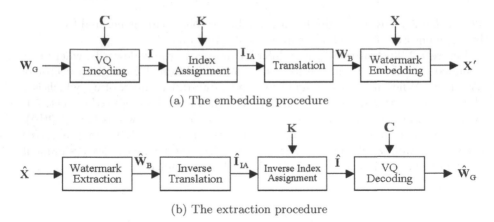

(a) The embedding procedure

(b) The extraction procedure

Fig. 7.3. The VQ-based watermarking scheme with a key for index assignment

7.2   Genetic Index Assignment (GIA)

Before executing the GIA procedure, the given gray watermark $\mathbf{W}_G$ is decomposed into T non-overlapping vectors of size d pixels. The VQ encoding procedure is performed using the assigned codebook $\mathbf{C}$. The indices of the nearest codewords obtained are collected as $\mathbf{I}$ and the steps below are done.

7.2.1   Steps

Step 1: Generate S keys $\{\mathbf{K}_1, \mathbf{K}_2, ..., \mathbf{K}_S\}$ randomly as the initial GA individuals.

Step 2: Convert $\mathbf{I}$ to $\{\mathbf{I}_{IA_1}, \mathbf{I}_{IA_2}, ..., \mathbf{I}_{IA_S}\}$ according to $\{\mathbf{K}_1, \mathbf{K}_2, ..., \mathbf{K}_S\}$.

Step 3: Translate $\{\mathbf{I}_{IA_1}, \mathbf{I}_{IA_2}, ..., \mathbf{I}_{IA_S}\}$ as binary bit streams $\{\mathbf{W}_{B_1}, \mathbf{W}_{B_2}, ..., \mathbf{W}_{B_S}\}$.

Step 4: Embed $\{\mathbf{W}_{B_1}, \mathbf{W}_{B_2}, ..., \mathbf{W}_{B_S}\}$ into $\mathbf{X}$ respectively to obtain S watermarked images $\{\mathbf{X}'_1, \mathbf{X}'_2, ..., \mathbf{X}'_S\}$.

Step 5: Evaluate the performance for all the individuals.

Step 6: Terminate the training procedure if the considered GA iteration is satisfied; otherwise, continue executing the following steps.

Step 7: Assign the probabilities to these individuals according to their performance.

Step 8: Regenerate S new individuals for next generation.

Step 9: Mutate the genes of these new individuals.

Step 10: Go to Step 2.

7.2.2   Details of the GIA Procedure

In Step 1, as the goal of the GIA procedure is to find a better way to assign indices to the codewords, the index-assignment key $\mathbf{K}$ therefore is regarded as an individual and its elements are regarded as its genes.

In Steps 2 and 3, the methods introduced in Sects. 7.1.1 and 7.1.2 are performed to achieve their aims. In Step 4, any of the suitable watermarking schemes proposed in the literature can be employed. Details of the embedding method is not the prime focus here.

In Step 5, in a similar way to the description of Sect. 6.4.1, the fitness function defined in Eq. (6.10) can be used to measure the performance of each individual. The fitness scores are then referred in Step 7 to assign the probabilities of being selected as parents for the production of offspring. The better performance the individual has, the higher probability is assigned to the individual.

The crossover step is performed to generate new individuals. The mutation procedure is done to mutate the genes of the new individuals according to the predefined mutation rate ρ_m. Details of these steps are similar to the steps described in Sects. 4.3.2 or 6.4.1.

The above steps are repeated until the considered iteration t is completed. Finally, the individual with the best performance is regarded as the index-assignment key $\mathbf{K}$ for the watermarking scheme mentioned in Sect. 7.1.3.

7.3   Simulation Results and Comparison

In the simulation, the images shown in Figs. 1.2 (LENA), 1.3 (PEPPERS), and 1.4 (BABOON) were used as the cover images. The size of either image is 512×512 pixels in gray-level. The image shown in Fig. 7.1(a) with size 256×256 pixels in gray-level was used as the original watermark. A codebook containing 16 codewords was obtained from Fig. 7.1(a) by applying the LBG algorithm [24] with a threshold of 0.0001. In the VQ-based gray watermark coding procedure, the gray watermark was decomposed into $T = 4096$ non-overlapping blocks of size 4×4 pixels and the codebook mentioned was used to encode it.

The VQ-based scheme presented in Sect. 6.3.3 was employed as the watermarking method. In this scheme, a codebook containing 256 codewords was obtained from the image of LENA by applying the LBG algorithm with a threshold of 0.0001. It was used to code all the cover images. The key for codebook partition was generated randomly without a training technique. For the GIA procedure, the below equation was used as the fitness function to evaluate the performance of each individual:

$$f_i = \mathrm{PSNR}(\mathbf{X}, \mathbf{X}'_i), \; i = 1, 2, ..., S. \tag{7.3}$$

Also, the settings of $S = 20$, $t = 1000$, and $\rho_m = 30\%$ were used.

When no GIA procedure was introduced, the encoded result (Fig. 7.1(c)) of the gray watermark was embedded into the respective cover images. When the GIA procedure was introduced, the encoded results (Fig. 7.4) of the gray watermark using the trained keys were embedded into the corresponding cover images. Table 7.1 lists the imperceptibility test results without using and using the GIA procedure.

To test the robustness, some common image processing procedures were employed to attack the watermarked images. Tables 7.2 and 7.3 list the test results without and with the GIA procedure. Figures 7.5 and 7.6 display the gray watermarks recovered from the attacked watermarked images when the GIA procedure is not used and when the GIA procedure is used.

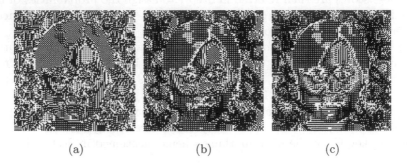

(a)   (b)   (c)

Fig. 7.4. The binary bit streams obtained, which are arranged as binary images of size 128 × 128 pixels, when the GIA procedure is introduced and the images of LENA, PEPPERS, and BABOON are used as the cover images respectively.

Table 7.1. Imperceptibility test results (PSNR) without and with GIA

Method	PSNR (dB)		
	LENA	PEPPERS	BABOON
VQ coding	31.800	28.028	22.537
No GIA	29.891	26.935	22.091
With GIA	30.150	27.131	22.150

Table 7.2. Robustness test results (BCR) when no GIA procedure is introduced

Attack	BCR (%)		
	LENA	PEPPERS	BABOON
VQ, L=512	97.760	98.413	98.981
JPEG, QF=60%	95.471	96.478	96.722
JPEG, QF=80%	99.713	99.774	99.762
Median filtering	92.883	90.973	81.934
Cropping, 25%	85.876	85.876	85.876

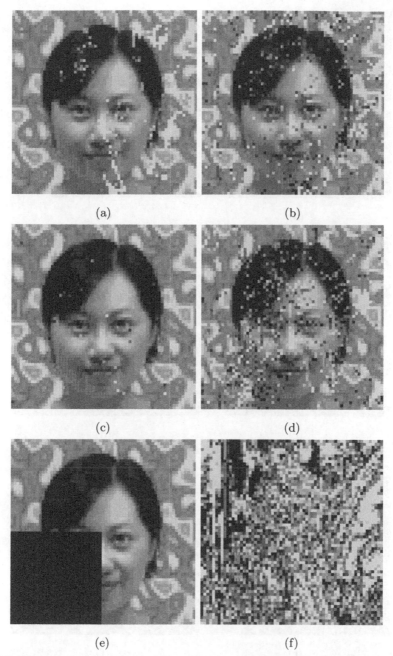

Fig. 7.5. The watermarks recovered from the attacked watermarked images while using LENA as the cover image and no GIA procedure is introduced: (a) VQ compression with a codebook size=512, (b) JPEG compression with a QF=60%, (c) JPEG compression with a QF=80%, (d) median filtering with a window size=3, (e) 25% of cropping in the left-bottom corner, (f) using the original LENA, which contains no watermark signal, as the input image for extraction.

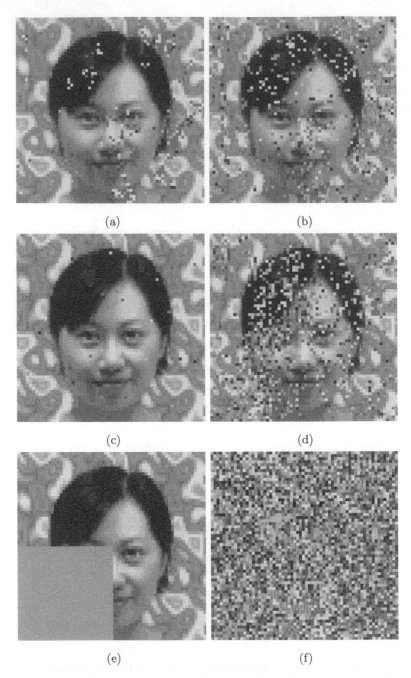

(a)   (b)

(c)   (d)

(e)   (f)

Fig. 7.6. The watermarks recovered from the attacked watermarked images while using LENA as the cover image and the GIA procedure is introduced: (a) VQ compression with a codebook size=512, (b) JPEG compression with a QF=60%, (c) JPEG compression with a QF=80%, (d) median filtering with a window size=3, (e) 25% of cropping in the left-bottom corner, (f) using a wrong codebook partition key for extraction.

Table 7.3. Robustness test results (BCR) when the GIA procedure is introduced

Attack	BCR (%)		
	LENA	PEPPERS	BABOON
VQ, L=512	97.003	96.858	98.444
JPEG, QF=60%	95.258	96.332	96.661
JPEG, QF=80%	99.799	99.744	99.573
Median filtering	93.127	92.413	80.853
Cropping, 25%	89.227	91.266	90.338

7.4 Summary

A genetic-index-assignment procedure based on vector quantisation for general watermarking schemes has been presented in this section. It applies GAs to find a better way to adjust the signal of the watermark to suit the signal of the given cover image. So that the imperceptibility can be improved. Introducing this concept in watermarking systems is novel, and the simulation results demonstrate its effectiveness.

Chapter 8
Genetic Watermark Modification

Two watermark modification methods – a lossless method and a lossy method – and a GA training procedure for improving the performance of general watermarking schemes are introduced in this chapter. These methods improve the imperceptibility and embedding capacity by modifying and compressing the original watermark according to the block table given and the indices assigned. Also, by introducing the GA training procedure, a better way to assign the indices according to the signal of the cover image can be found. This therefore provides better imperceptibility.

8.1 Watermark Modification

Modifying the signal of the input watermark to have better watermarking result, is the main concept of watermark modification. In this section, a lossless watermark modification method and a lossy watermark modification method proposed in [98] are introduced. Both methods modify the signal of the original watermark $\mathbf{W}$ according to a user-defined index table $\mathbf{K}$, as shown in Fig. 8.1. The modified watermark $\mathbf{B}$ is embedded into the given cover image $\mathbf{X}$ to generate a watermarked image $\mathbf{X}'$. In these watermark modification methods, any suitable watermarking methods proposed in the literature can be applied to implement the embedding and the extraction steps.

8.1.1 Lossless Method

To modify a given binary watermark $\mathbf{W}$, this watermark is first decomposed into T non-overlapping blocks $\{\mathbf{w}_1, \mathbf{w}_2, ..., \mathbf{w}_T\}$ of size m pixels. Here the block can be in one-dimension or two-dimension format. A block set $\mathbf{V}$ which contains all the possible types $p = 2^m$ of blocks can then be formed. For example, Table 8.1 lists all the possible types of blocks when $m = 2 \times 2$ pixels. According to a given index table $\mathbf{K}$ which is a set of p positive integers $\{k_1, k_2, ..., k_p\}$, where for any integers $i : 1 \leq i \leq p$ and $j : 1 \leq j \leq p$, $1 \leq k_i \leq p$ and $k_i \neq k_j$ if $i \neq j$, each index of $\mathbf{K}$ is assigned to the corresponding type of block. That is, k_1 is assigned to the type-1 block in Table 8.1.

F.-H. Wang, J.-S. Pan, and L.C. Jain: Innovations in Dig. Watermark. Tech., SCI 232, pp. 119–127.
springerlink.com   © Springer-Verlag Berlin Heidelberg 2009

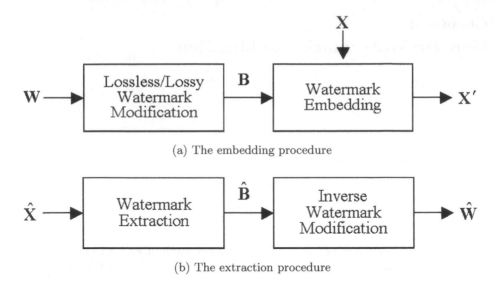

(a) The embedding procedure

(b) The extraction procedure

Fig. 8.1. The brief block diagrams of the proposed lossless/lossy watermark modification method for general watermarking methods

Table 8.1. All possible types of blocks when $m = 2 \times 2$ pixels

Type	Block	Content	Assigned Index
1		0000	k_1
2		0001	k_2
3		0010	k_3
4		0011	k_4
5		0100	k_5
6		0101	k_6
7		0110	k_7
8		0111	k_8
9		1000	k_9
10		1001	k_{10}
11		1010	k_{11}
12		1011	k_{12}
13		1100	k_{13}
14		1101	k_{14}
15		1110	k_{15}
16		1111	k_{16}

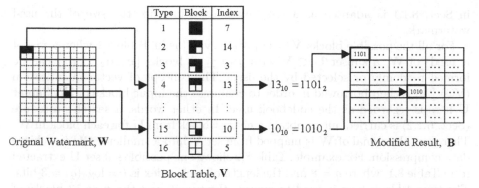

Original Watermark, **W**

Block Table, **V**

$13_{10} = 1101_2$

$10_{10} = 1010_2$

Modified Result, **B**

Fig. 8.2. Illustration of generating the output bit stream from the original watermark

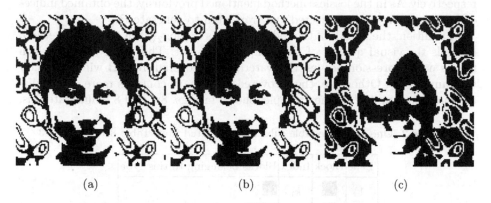

(a)   (b)   (c)

Fig. 8.3. Examples of modifying the original watermark using different index tables when Table 8.1 is used as block set **V**: (a) the original watermark, (b) the modified result when $\{k_1, k_2, ..., k_p\} = \{0, 1, ..., 15\}$, and (c) the modified result when $\{k_1, k_2, ..., k_p\} = \{15, 14, ..., 0\}$.

Afterwards, by use of a search procedure, the block types of $\{\mathbf{w}_1, \mathbf{w}_2, ..., \mathbf{w}_T\}$ and the indices associated to the types can be obtained. We then translate those indices from integer format into binary format to form an output bit stream **B** as the method described in Sect. 7.1.1. This bit stream is the adjusted signal of **W**. The above modification process using Table 8.1 is depicted in Fig. 8.2. Two examples of the modified results using Table 8.1 and different index tables are given in Fig. 8.3.

To recover the original watermark from **B**, the inverse steps of the above procedure are carried out merely.

8.1.2 Lossy Method

To expand the embedding capacity of a watermarking system, compressing the input watermark into smaller size is a sensible solution. The method introduced in Sect. 8.1.1 is adapted as a lossy method to increase the size of the used watermark.

For all the possible blocks $\mathbf{V} = \{\mathbf{v}_1, \mathbf{v}_2, ..., \mathbf{v}_p\}$ (as in Table 8.1 when a $m = 4$ is used) of $\mathbf{W}$, a subset $\mathbf{U} \subset \mathbf{V}$ with q ($q \leq p$) blocks $\{\mathbf{u}_1, \mathbf{u}_2, ..., \mathbf{u}_q\}$ can be formed randomly or selected by the users. The concept of vector quantisation (VQ) [24] is then borrowed in order to compress the input blocks of $\mathbf{W}$. Subset $\mathbf{U}$ here is regarded as the codebook used. In other words, a search procedure (Sect. 6.2.2) is carried out to obtain a nearest block from $\mathbf{U}$ for each block in $\mathbf{W}$. Therefore, the signal of $\mathbf{W}$ is mapped from set $\mathbf{V}$ into a smaller set $\mathbf{U}$ to achieve data compression. For example, Table 8.2 shows one possible subset $\mathbf{U}$ extracted from Table 8.1, where $q = 8$ and the length of each index is $l = \log_2(q) = 3$ bits. The type-4 block here is used to express the type-6 and the type-13 blocks of Table 8.1. Other examples of $l = 2$ and $l = 1$ are given in Tables 8.3 and 8.4 respectively. As in the lossless method mentioned previously, the obtained indices of $\{\mathbf{w}_1, \mathbf{w}_2, ..., \mathbf{w}_T\}$ are collected and translated to generate $\mathbf{B}$.

In addition, the length of the index, which is $l = \log_2(p)$ in the lossy method, controls the visual quality of the compressed result. That is, if $l = m$, which means no compression occurs, the watermark recovered from $\mathbf{B}$ will be exactly the same as $\mathbf{W}$. Otherwise, the recovered watermark contains more or less distortion according to l. Some examples shown in Figs. 8.4 and 8.5 illustrate this,

Table 8.2. One of the possible subset $\mathbf{U}$ extracted from Table 8.1 when $l = 3$

Type	Block	Index	Blocks belonging to this type
1	■	k_1	■
2	▛	k_2	 ▟ ▛
3	▙	k_3	▙ ▟
4	▘	k_4	 ▛ ▜ ▟
5	▚	k_5	▛ ▞ ▟
6	�

k_6	▟ ▛
7	▗	k_7	 ▟ ▛
8	▦	k_8	▦

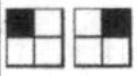

Table 8.3. One of the possible subset $\mathbf{U}$ extracted from Table 8.1 when $l = 2$

Type	Block	Index	Blocks belonging to this type
1	■	k_1	 ■ ▛ ▜ ▙ ▟
2	▚	k_2	 ▚ ▛ ▞ ▜ ▟
3	▘	k_3	 ▛ ▜ ▟ ▛
4	▦	k_4	▦

Table 8.4. One of the possible subset **U** extracted from Table 8.1 when $l = 1$

Type	Block	Index	Blocks belonging to this type
1	■	k_1	■ ◪ ◪ ◪ ◪ ◪ ◪
2	⊞	k_2	⊞ ⊞ ⊞ ⊞ ⊞ ⊞ ⊞ ⊞ ⊞

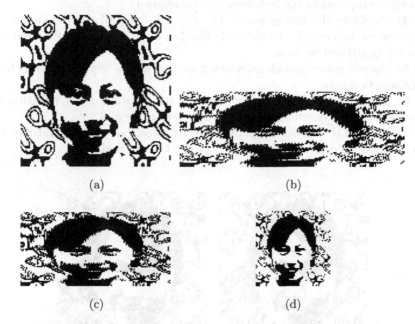

(a)   (b)

(c)   (d)

Fig. 8.4. Examples of the modified results, which are arranged as binary images for display, when Fig. 8.3(a) is used as the original watermark: (a) $l = 4$ and Table 8.1 is used, (b) $l = 3$ and Table 8.2 is used, (c) $l = 2$ and Table 8.3 is used, and (d) $l = 1$ and Table 8.4 is used.

where Table 8.1 with $\mathbf{K} = \{0, ..., 15\}$, Table 8.2 with $\mathbf{K} = \{0, ..., 7\}$, Table 8.3 with $\mathbf{K} = \{0, ..., 3\}$, and Table 8.4 with $\mathbf{K} = \{0, 1\}$ are used to encode Fig. 8.3(a) respectively. According to the values of the l used in the corresponding subsets, Fig. 8.3(a) here is compressed into 4/4, 3/4, 1/2, and 1/4 of its original size, as shown in Fig. 8.4. Clearly, embedding fewer watermark bits into the cover image helps watermarking systems provide better imperceptibility.

8.2 Genetic Watermark Modification (GWM)

Considering how to find a better way to assign indices to the blocks of **V**, a genetic watermark modification (GWM) procedure [98] is developed and described here.

Step 1: Generate S index tables $\{\mathbf{K}_1, \mathbf{K}_2, ..., \mathbf{K}_S\}$ randomly as the initial GA individuals.

Step 2: Obtain S watermark bit streams $\{\mathbf{B}_1, \mathbf{B}_2, ..., \mathbf{B}_S\}$ by employing either of the methods introduced in Sect. 8.1 according to $\{\mathbf{K}_1, \mathbf{K}_2, ..., \mathbf{K}_S\}$, respectively.

Step 3: Embed $\{\mathbf{B}_1, \mathbf{B}_2, ..., \mathbf{B}_S\}$ into the cover image $\mathbf{X}$ respectively to form S watermarked images $\{\mathbf{X}'_1, \mathbf{X}'_2, ..., \mathbf{X}'_S\}$. Here any suitable watermarking method proposed in the literature can be applied.

Step 4: Calculate the fitness scores $\{f_1, f_2, ..., f_S\}$ for all the individuals according to the embedded results obtained in Step 3. A fitness function such as Eq. (6.10) can be used.

Step 5: Discard some individuals which have bad fitness scores according to ρ_s, the selection rate.

Step 6: Terminate the training procedure if the considered GA iteration t is met or continue executing the following steps otherwise.

Step 7: Perform the crossover operation to produce S new individuals for next generation, as described in the GPS or the GIA procedures.

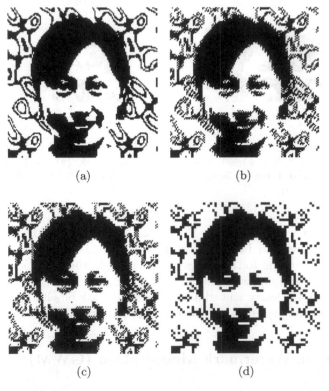

(a)   (b)

(c)   (d)

Fig. 8.5. Examples of recovering watermarks from Fig. 8.4, where (a) Table 8.1 is used, (b) Table 8.2 is used, (c) Table 8.3 is used, and (d) Table 8.4 is used

Step 8: Mutate the genes of the new individuals according to a predefined mutation rate ρ_m.

Step 9: Go to Step 2.

Details of the steps listed above are similar to the steps described in the GPS (see Sect. 4.3), GCP (see Sect. 6.4), and GIA (see Sect. 7.2) procedures.

8.3 Simulation Results and Comparison

In our experiments, the image of LENA (512 × 512 pixels in gray-level, Fig. 1.2) was used as the cover image. The watermarking systems proposed in [92], [77], and [93], which are the spatial-based scheme, the transform-based scheme, and the VQ-based scheme, respectively, were employed as the test watermarking systems. For [92], the pixels selected from each 8 × 8 block for embedding were (18, 29, 42, 53) and the delta used was 10. For [77], the bands used in each DCT block for embedding were (14, 15, 16, 27). For [93], the codebook size was 256 and the codebook-partition key was generated randomly. The fitness function used in the GWM procedure was

$$f_i = \text{PSNR}(\mathbf{X}, \mathbf{X}'_i), \ i = 1, 2, ..., S. \tag{8.1}$$

The settings for the GWM procedure were: $S = 10$, $\rho_s = 0.5$, $\rho_m = 0.3$, and $t = 200$.

8.3.1 Results of Lossless Method

In the experiments of the lossless watermark modification, the image shown in Fig. 8.3(a) with size 128 × 128 pixels in bi-level was used as the original watermark. We segmented it into 4096 blocks ($m = 4$), 2048 blocks ($m = 8$), and 1024 blocks ($m = 16$), respectively. Results of these experiments are listed in Table 8.5.

Table 8.5. Embedded results of the lossless watermark modification method

Watermarking scheme	PSNR (dB)			
	No GWM	With GWM		
		$m = 4$	$m = 8$	$m = 16$
Spatial-based [92]	38.811	38.976	39.005	39.017
Transform-based [77]	30.113	32.875	32.235	31.113
VQ-based [93]	30.055	30.211	30.203	30.191

8.3.2   Results of Lossy Method

In the experiments of the lossy GWM method, to show that by controlling the value of l embedding different sizes of watermarks becomes possible, the gray watermarks shown in Fig. 8.6 were used instead. The sizes of them are 256×256 pixels and 128×128 pixels respectively. Two subsets which contain 16 blocks ($m = 4 \times 4$ pixels and $l = 4$ bits) and 256 blocks ($m = 4 \times 4$ pixels and $l = 8$ bits) were obtained from Fig. 8.6 by employing the LBG algorithm [24]. They were used to compress Fig. 8.6(a) and Fig. 8.6(b) respectively. Table 8.6 lists the sizes of the original watermarks and the encoded results. Tables 8.7 and 8.8 show the test results of the lossy modification method.

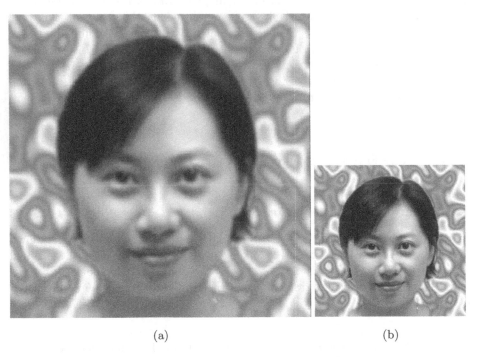

(a)   (b)

Fig. 8.6. The watermarks used in the lossy GWM experiments

Table 8.6. The comparison of the encoded results while applying the lossy watermark modification method to Fig. 8.6

Watermark	Original size (bits)	Encoded size (bits)
Fig. 8.6(a)	524,288 ($256 \times 256 \times 8$)	16,384 ($64 \times 64 \times 4$)
Fig. 8.6(b)	131,072 ($128 \times 128 \times 8$)	8,192 ($32 \times 32 \times 8$)

Table 8.7. Embedded results (PSNR) of the lossy GWM method when a $m = 4 \times 4$ pixels, a $l = 4$, and Fig. 8.6(a) are used

Watermarking scheme	PSNR (dB)	
	No GWM	With GWM
Spatial-based [92]	38.839	38.982
Transform-based [77]	30.327	31.409
VQ-based [93]	30.050	30.193

Table 8.8. Embedded results (PSNR) of the lossy GWM method when a $m = 4 \times 4$ pixels, a $l = 8$, and Fig. 8.6(b) are used

Watermarking scheme	PSNR (dB)	
	No GWM	With GWM
Spatial-based [92]	42.151	42.365
Transform-based [77]	32.331	32.544
VQ-based [93]	30.919	31.092

8.4 Discussions and Summary

The watermark modification methods introduced show a way to improve imperceptibility by employing the intelligent technique in the watermarking systems. The lossy watermark modification method also shows a way to expand the capacity of the embedded bits, as the results shown in Table 8.6. Again, the experimental results listed demonstrate how the watermarking systems can be improved by employing intelligent techniques, which is the GWM methods introduced. The robustness of the watermarking systems with or without the GWM procedure generally depends on the characteristics of the corresponding watermarking systems. This and the key delivery issue are beyond the scope of this chapter.

Part III
Hybrid Systems of Digital Watermarking

Chapter 9
Watermarking Based on Multiple Description VQ

Considering transferring watermarked data over channels, we may encounter one question: can the signals of the embedded watermark(s) survived under the noise of the channels? In this chapter, unlike the watermarking systems introduced previously which apply artificial procedures (or attacks in other term) to test the robustness, we focus on the channel noise and introduce a hybrid watermarking scheme to overcome the losses caused by channel errors. The system is based on VQ and multiple description coding (MDC), and also has the ability of embedding dual watermarks. The VQ-based watermarking system first modifies the VQ indices to hide the first watermark, as illustrated in Chap. 6, and applies the well-known traditional LSB scheme to hide the second watermark, as introduced in Chap. 4. The concept of MDC is then employed to generate the considered number of descriptions, which are transmitted to the receivers via different channels. On the receiver side, accoring to the number of descriptions received, the embedded dual watermarks can be extracted by applying the corresponding inverse embedding procedures. At the end of this chapter, we present the simulation results to highlight the superiority of this scheme.

9.1 Multiple Description Coding (MDC)

During transmissions of data, losses are inevitable due to channel errors or lost packets under different types of transmission channels. In contrast with the conventional schemes such as progressive transmission, multiple description coding (MDC) offers another scope for effective transmission of compressed multimedia information.

9.1.1 Introduction

MDC is an error resilient coding technique, which can be cast as a source coding method for a channel whose end-to-end performance includes uncorrected erasures. Applications of MDC focus on error concealment and error resilience. This channel is encountered in a packet communication system that has effective error

F.-H. Wang, J.-S. Pan, and L.C. Jain: Innovations in Dig. Watermark. Tech., SCI 232, pp. 131–149.
springerlink.com   © Springer-Verlag Berlin Heidelberg 2009

detection but does not have retransmission of incorrect or lost packets. MDC uses diversity to overcome channel impairments such that a decoder, which receives an arbitrary subset of the channels, may re-produce a useful reconstruction [22]. Information-theoretic issues of MDC have been studied extensively since early eighties [23] [108]. In multiple description (MD) coders, the same source material is coded into several chunks of data, called *descriptions*, such that each description can be decoded independently to obtain a minimum fidelity; while combining with other descriptions to achieve the better quality. By doing so, although the goals for MDC and channel coding are to make effective transmission of data, MDC offers a totally different perspective from channel coding [50].

For transmission, MDC is suitable for noisy channels with long bursts of errors. To gain robustness of the loss of descriptions, MDC must sacrifice some compression efficiency while retaining the capability of error resilience. Figure 9.1 depicts the generic model for MD source coding with two channels and three decoders. The input source $\mathbf{x}$, for example, an image or vector, is encoded and two descriptions d_1 and d_2 are generated. Then, each of the descriptions is transmitted to the decoder side via different channels respectively. For instance, in Fig. 9.1 d_1 is sent via Channel 1, and d_2 is sent via Channel 2. On the decoder side, after receiving the descriptions d_1' and d_2', the decoders reconstruct the input source and output $\mathbf{x}'^{(0)}$, $\mathbf{x}'^{(1)}$, and $\mathbf{x}'^{(2)}$.

In the MDC system, Decoder 0 usually is called the *central decoder*, and Decoders 1 and 2 are called the *side decoders*. The Euclidean distance between $\mathbf{x}$ and $\mathbf{x}'^{(0)}$ is the *central distortions*, while the errors between $\mathbf{x}$ and $\mathbf{x}^{(i)}$, $i = 1, 2$, are the *side distortions*. It suggests a situation in which there are three separate users or three classes of users, which could arise in broadcasting on two channels. The same abstraction holds if there is a single user that can be in one of three states depending on which descriptions are received. Generally speaking, if we extend the number of transmission channels in Fig. 9.1 to K, there will be $2^K - 1$ receivers that decode with different number of descriptions received, and obtain different qualities of the reconstructed image.

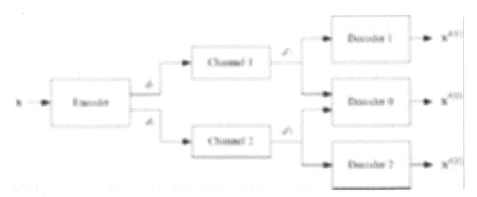

Fig. 9.1. The generic model for MD source coding with two channels and three receivers. The general case has K channels and $2^K - 1$ receivers.

In addition to making theoretic researches, it is also important to devise practical designs to make MDC applicable with the scenario depicted in Fig 9.1. Practical applications and implementations of MDC emerged in the nineties. Two major categories for MDC applications are: (i) *quantisation based* schemes, such as multiple description scalar quantisation (MDSQ) [84] and multiple description vector quantisation (MDVQ) [26], and (ii) *transform-domain based* schemes, called multiple description transform coding (MDTC) [103] [104].

In this chapter, we focus on quantisation based MD schemes for watermarking. Figures 9.2 and 9.3 illustrate the structure of MDSQ and MDVQ respectively for two descriptions over two independent channels with mutually independent breakdown probabilities. And, between these two systems, MDVQ will be the focused one. We will describe the details of it in the following sections.

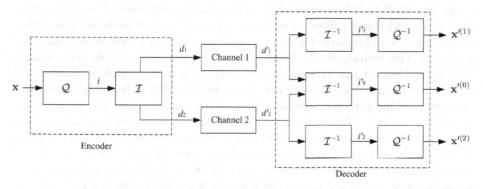

Fig. 9.2. The structure of MDSQ for two descriptions over two independent channels with mutually independent breakdown probabilities.

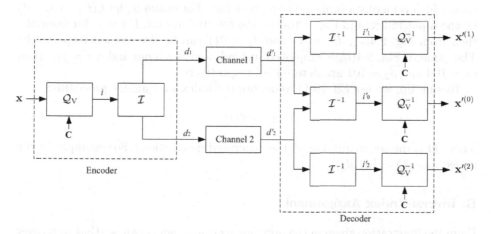

Fig. 9.3. The structure of MDVQ for two descriptions over two independent channels with mutually independent breakdown probabilities

9.1.2   Index Assignment and Inverse Index Assignment

Before going further, we introduce the index assignment procedure in this section.

A. Index Assignment

As shown in Fig. 9.2 or Fig. 9.3, there is a step $\mathcal{I}$ named "index assignment" used in the MD decoder. This step accepts an index i and produces a set of descriptions $\{d_1, d_2, ..., d_K\}$. Here K is the number of descriptions or the number of channels. For simplicity, we set $K = 2$ in this chapter. There are many methods proposed in the literature for the purpose of index assignment. Here the one introduced in [84] is described.

As described in [84], optimization of the index assignment $\mathcal{I}$ is not easy, so instead of addressing the exact optimal index assignment problem, the author suggested several heuristic techniques, and one of them is the *nested index assignment*. This technique may likely give close to the best possible performance. The basic ideas of this technique is to prepare a matrix, and to number the cells of the matrix from upper-left to lower-right by filling from the main diagonal outward increasingly. The dimension of the matrix is denoted by K, which equals the number of descriptions or the number of channels for transmission. For $K = 2$, the descriptions of the MDC can be interpreted as the row and column indices of the matrix. The author considered a set of index pairs constructed from those that lie on the main diagonal and on the $2k$ diagonals closest to the main diagonal, where the parameter k is called *spread*. Figure 9.4 gives two examples for addressing how the technique functions. Here $K = 2$ and codebook size $L = 8$, which equals the number of column or row of the matrix, are used.

To use the index assignmet matrix to produce a set of K descriptions is very easy, which is merely a table look-up procedure. For example, let i ($i \in [0, L-1]$) be the input index, and Fig. 9.4(a) be the referred matrix. If $i = j_3$ for example, then from Fig. 9.4(a), $d_1 = 011$ and $d_2 = 011$ can be determined respectively. The same, if Fig. 9.4(b) is employed instead, and the input index $i = j_{13}$, then $d_1 = 100$ and $d_2 = 101$ are determined respectively.

To sum up, we use Eq. (9.1) to denote the index assignment procedure:

$$\mathbf{D} = \mathcal{I}(i), \tag{9.1}$$

where $\mathbf{D}$ is the set containing all the generated descriptions. For example, in the above case, $\mathbf{D} = \{d_1, d_2\}$.

B. Inverse Index Assignment

From the illustration given in the previous section, readers can see that to recover the original index from a set of descriptions is not difficult. The same, the table look-up procedure is applyed merely. For example, let $d_1 = 100$ and $d_2 = 101$ be the output results of the index assignment procedure, and Fig. 9.4(a) be

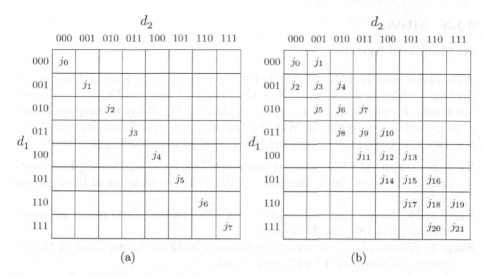

Fig. 9.4. The illustrations of the nested index assignment in MDSQ for two channels with the codebook size $L = 8$. (a) With spread $k = 0$. (b) With spread $k = 1$.

the referred matrix. To recover the original index i from d_1 and d_2, the row and column indicated by d_1 and d_2 are picked respectively. The intersection of Column d_1 and Row d_2 of Fig. 9.4(a) is regarded as i. That is $i = j_3$.

However, as we mentioned previously, MDC is an error resilient coding technique, which implies that suffering from transimission loss must be very common. It means that some of the descriptions used for recovering the original index may not be received. For instance, in the above example the decoder may only receives d_1. Thus, how to use the received descriptions to determine what the original index is should be considered. The inverse index assignment is adapted as:

$$\mathbf{I} = \mathcal{I}^{-1}(\mathbf{D}') \, . \tag{9.2}$$

Here $\mathbf{D}'$ is the set containing all the received descriptions, $\mathbf{I}$ is the set containing all the possible recovered indices, and $\mathcal{I}^{-1}$ denotes the inverse index assignment procedure. If $\mathbf{D}'$ contains all the descriptions sent by the encoder (that is $\mathbf{D}' = \mathbf{D}$), then Eq. (9.2) will become:

$$i = \mathcal{I}^{-1}(\mathbf{D}) \, . \tag{9.3}$$

To let readers understand easily, we give some examples here. Let $d_1 = 100$ and $d_2 = 101$ be the generated descriptions of the encoder, and Fig. 9.4(a) be the referred matrix. If only d_1 is received on the decoder side, then Column d_1 of Fig. 9.4(a) is picked. As we can see, j_3 is the only possible index in this column, so we set $i = j_3$. In this case, $\mathbf{D}' = \{d_1\}$ and $\mathbf{I} = \{j_3\}$. Now, in the above example, if Fig. 9.4(b) is used as the referred matrix instead, then we have: Column d_1 of Fig. 9.4(b) is selected, and the indices containing in this column, which are j_{11}, j_{12}, and j_{13}, are the possible candiates of i. So, $\mathbf{I} = \{j_{11}, j_{12}, j_{13}\}$.

9.1.3   MDVQ

A. Simplized System

A simplized version of MDVQ based on $k = 0$ and $K = 2$ is depitched in Fig. 9.3. To begin with, we denote that $\mathbf{X}$ is the original image, and $\mathbf{C}$ is a codebook containing L codewords $\{c_0, c_1, ..., c_{L-1}\}$ therein. We firstly decompose $\mathbf{X}$ into T non-overlapping blocks (or vectors) $\{\mathbf{x}_0, \mathbf{x}_1, ..., \mathbf{x}_{T-1}\}$. Then, for each block $\mathbf{x} \in \{\mathbf{x}_0, \mathbf{x}_1, ..., \mathbf{x}_{T-1}\}$, the steps below are used as the encoding procedure:

Step 1: Execute the VQ encoding procedure (see Sect. 6.2.2) to find a nearest codeword from $\mathbf{C}$ for $\mathbf{x}$. Here let $i \in [0, L-1]$ be the index of the obtained codeword.

Step 2: Do index assignment to produce a pair of descriptions from i by applying Eq. (9.1). We have $\mathbf{D} = \{d_1, d_2\}$.

Step 3: Afterwards, the descriptions colleced in $\mathbf{D}$ are transmitted to the receivers via Channels 1 and 2, respectively.

On the decoder side, after receiving the descriptions from Channels 1 and 2, the descriptions are collected as $\mathbf{D}'$. The below procedure is then carried out to reconstruct the output image:

Step 1: Apply the inverse index assignment procedure (Eq. (9.1)) to recover the original index from $\mathbf{D}'$. For example, for Decoder 0, the descriptions received are d_1' and d_2'; for Decoder 1, the received is d_1' merely; for Decoder 2, d_2' is the received description. Here we use i' to denote the output result of the inverse index assignment procedure.

Step 2: Output the codeword whose index is i' from $\mathbf{C}$. This codeword is regarded as the reconstructed block. That is, for Decoder 0, $\mathbf{x}'^{(0)} = \mathbf{c}_{i'}$; for Decoder 1, $\mathbf{x}'^{(1)} = \mathbf{c}_{i'}$; for Decoder 2, $\mathbf{x}'^{(2)} = \mathbf{c}_{i'}$.

For each decoder, after all the blocks are reconstructed from the received descriptions, they are pieced together to form an output image. That is, $\mathbf{X}'^{(0)} = \{\mathbf{x}_0'^{(0)}, \mathbf{x}_1'^{(0)}, ..., \mathbf{x}_{T-1}'^{(0)}\}$, $\mathbf{X}'^{(1)} = \{\mathbf{x}_0'^{(1)}, \mathbf{x}_1'^{(1)}, ..., \mathbf{x}_{T-1}'^{(1)}\}$, and $\mathbf{X}'^{(2)} = \{\mathbf{x}_0'^{(2)}, \mathbf{x}_1'^{(2)}, ..., \mathbf{x}_{T-1}'^{(2)}\}$.

B. Normalized System

In the following sections, as we are not concerned which decoder on the decoder side should be used to reconstruct the output image, the MDVQ system can then be normalized as Fig. 9.5.

In Fig. 9.5, the definitions of $\mathbf{x}$, $\mathbf{C}$, i, d_1, d_2, d_1', d_2', and $\mathbf{x}'$ are all the same as the definitions of that in Fig. 9.3. $\mathbf{D}$ denotes the set containing all the K descriptions generated by $\mathcal{I}$. $\mathcal{D}$ denotes the distribution procedure which sends each description in $\mathbf{D}$ to the decoder side via the corresponding channel. $\mathcal{C}$ denotes the collection procedure, which collects all the received descriptions as $\mathbf{D}'$.

Here if no channel is broken down, both descriptions d_1' and d_2' can be received. In this case, $\mathbf{D}' = \{d_1', d_2'\}$, and the behavior of the decoder can be regarded

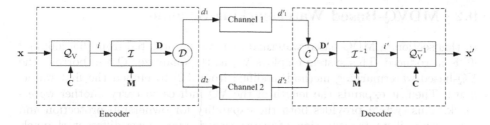

Fig. 9.5. The structure of the normalized MDVQ system for $k = 0$ and $K = 2$

as Decoder 0 of Fig. 9.3. If Channel 1 is broken down, then only d'_2 can be received. In this case, $\mathbf{D}' = \{d'_2\}$ and the decoder works as Decoder 2 of Fig. 9.3. Identically, the decoder works as Decoder 1 shown in Fig. 9.3 while Channel 2 is broken.

Now, to consider the cases of $k \geq 0$ and $K \geq 2$, the system presented in Fig. 9.5 should be extended. Figure 9.6 is the entended model. Here due to the behavior of the encoder shown in Fig. 9.5 is exactly the same to that of Fig. 9.6, we therefore only focus on the details of the decoder.

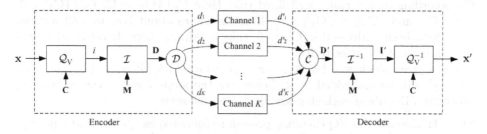

Fig. 9.6. The structure of the normalized MDVQ system for $k \geq 0$ and $K \geq 2$

As addressed in the inverse index assignment procedure, while $k > 0$ and some of the sent descriptions are missing, that is $||\mathbf{D}|| > ||\mathbf{D}'||$, the inverse index assignment procedure $\mathcal{I}^{-1}$ will not be able to tell what the originl index is. Instead, it will tell what possible candicates are. In this case, all the possible candicates are collected as $\mathbf{I}'$. We use $N = ||\mathbf{I}'||$ to denote the number of indices contained in $\mathbf{I}'$. Then, the inverse VQ (denoted as $\mathcal{Q}_V^{-1}$) is executed for N times. Each time the index in $\mathbf{I}'$ is fetched, and the codeword with this index is output from $\mathbf{C}$ and regarded as $\mathbf{x}'$. For example, if $\mathbf{I}' = \{2, 6, 7\}$, then the 2nd, 6th, and 7th codewords of $\mathbf{C}$ are output respectively.

After getting all the N reconstructed blocks ready, we have $\{\mathbf{x}'_1, \mathbf{x}'_2, ..., \mathbf{x}'_N\}$. We calculate the average value for each corresponding pixel containing in each block, and have the finial reconstructed block $\mathbf{x}'$.

$$\mathbf{x}'[i] = \frac{\sum_{j=1}^{N} \mathbf{x}'_j[i]}{N} \, , i = 0, 1, ..., S, \qquad (9.4)$$

where $S = ||\mathbf{x}'||$ denotes the size of the reconstructed block.

9.2 MDVQ-Based Watermarking Scheme

In this section, a MDC-based watermarking system adapted from [65] [66] and [9] is introduced. This system employs VQ as the quantizer Q and borrows the VQ-based watermarking method modified from [52] to embed the first watermark. Then, it expands the length of the VQ indices to carry another watermark. This system provides both the capability for ownership protection and the error-resilient transmission of watermarked image over different channels with independent breakdown probabilities.

9.2.1 Watermark Embedding

The structure of the embedding procedure is illustrated in Fig. 9.7. Here $\mathcal{E}_1$ denotes the step for embedding the first watermark bit, and $\mathcal{E}_2$ denotes the step for embedding the second watermark bit. Other sybmols used have the same definitions as illustrated previously.

For a given watermark $\mathbf{W}_1$ consisting of N bits $\{w_{1,0}, w_{1,1}, ..., w_{1,N-1}\}$ and a cover image $\mathbf{X}$, we first decompose $\mathbf{X}$ into N non-overlapping blocks $\{\mathbf{x}_0, \mathbf{x}_1, ..., \mathbf{x}_{N-1}\}$. Then, the codebook $\mathbf{C}$ is split into two subcodebooks $\mathbf{C}_0$ and $\mathbf{C}_1$ according to a certain perdefined rule. Here $\mathbf{C}_0 \bigcup \mathbf{C}_1 = \mathbf{C}$, $\mathbf{C}_0 \bigcap \mathbf{C}_1 = \emptyset$, $L = ||\mathbf{C}||$, and $||\mathbf{C}_0|| = ||\mathbf{C}_1|| = \frac{L}{2}$. Here the issues about how to split a codebook have been addressed in Chap. 6, thus is skipped here. Readers who have the interests please refer to it.

For a given block $\mathbf{x} \in \{\mathbf{x}_0, \mathbf{x}_1, ..., \mathbf{x}_{N-1}\}$, watermark bit $w_1 \in \{w_{1,0}, w_{1,1}, ..., w_{1,N-1}\}$ will be embedded in $\mathbf{x}$. Below are the steps used as the embedding procedure. Details of each step are illustrated followed.

Step 1: Execute the VQ encoding procedure (denoted as Q_V) to find a nearest codeword from $\mathbf{C}$ for $\mathbf{x}$. The index of this codeword, which is denoted as i, is output to the next step.

Step 2: Do the first embedding procedure $\mathcal{E}_1$ to modify the value of i, so that w_1 can be carried. We denote i_S, which indicates the index contains a single watermark bit, as the output index of this step. Details of this step is explained below.

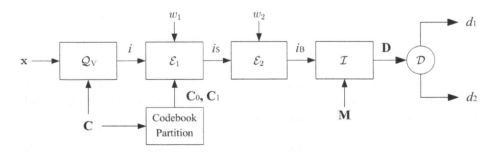

Fig. 9.7. The structure for watermarking two watermarks with two descriptions for transmission in MDC

Step 3: Apply the second embedding procedure $\mathcal{E}_2$ to expand the length of i_S, so that w_2 can be hidden within it. We use i_B, which indicates both watermark bits are contained within the index, to denote the output result. The same, details of this step are given and illustrated in the following section.

Step 4: Run the index assignemt procedure $\mathcal{I}$ to generate K descriptions, which are collected as **D**.

Step 5: Finally, execute the distributing procedure $\mathcal{D}$ to send all the descriptions to the receiver side via different memoryless and mutually independent channels respectively.

A. Embedding First Watermark

To let the index i carry the watermark bit w_1 by updating the content of i, either method introduced in Chap. 6 or in the literature can be employed. Here the example of the method modified from [52] is given.

Let codebook $\mathbf{C} = \{\mathbf{c}_0, \mathbf{c}_1, ..., \mathbf{c}_9\}$ containing $L = 10$ codewords. After partitioning $\mathbf{C}$ into two subcodebooks $\mathbf{C}_0$ and $\mathbf{C}_1$, assume we have $\mathbf{C}_0 = \{\mathbf{c}_9, \mathbf{c}_3, \mathbf{c}_4, \mathbf{c}_6, \mathbf{c}_0\}$ and $\mathbf{C}_1 = \{\mathbf{c}_7, \mathbf{c}_2, \mathbf{c}_5, \mathbf{c}_8, \mathbf{c}_1\}$. Here we can see that $\mathbf{C}_0 \bigcup \mathbf{C}_1 = \mathbf{C}$, $\mathbf{C}_0 \bigcap \mathbf{C}_1 = \emptyset$, and $||\mathbf{C}_0|| = ||\mathbf{C}_1|| = \frac{L}{2} = 5$. Then, for the given block $\mathbf{x}$, if $\mathbf{c}_6$ is the nearest codeword to it, we can know easily that $\mathbf{c}_6$ is belonging to $\mathbf{C}_0$, and it is the 4th codeword of $\mathbf{C}_0$. Now, to embed w_1 into $\mathbf{x}$, the 4th codeword should be picked from eitehr $\mathbf{C}_0$ or $\mathbf{C}_1$. That is, the 4th codeword of $\mathbf{C}_{w_1}$ is selected. If $w_1 = 1$, then $\mathbf{C}_{w_1} = \mathbf{C}_1$ is used and its 4th codeword, which is $\mathbf{c}_8$, is picked. Otherwise, $\mathbf{C}_0$ is used and its 4th codeword, which is $\mathbf{c}_6$, is selected. After determining which codeword should be picked, we go back to check $\mathbf{C}$ and fetch the index the codeword has. That is, if $\mathbf{c}_8$ is picked, then we output its index, which is 8 in this example, and set $i_S = 8$.

Please note, there are some requirements to be met when splitting $\mathbf{C}$ as $\mathbf{C}_0$ and $\mathbf{C}_1$. Randomly splitting $\mathbf{C}$ usually results in poor visual quality of image reconstruction. Also, the above method for determining which codeword should be used is not so effective in some aspects. For instance, the time used for obtaining the output codeword and the distortion between $\mathbf{x}$ and the picked codeword may be increased. We encourage the reader who has the interest to read Chap. 6 for further details.

B. Embedding Second Watermark

To embed another watermark bit into i_S is not difficult. Let $w_2 \in \mathbf{W}_2$ be the bit to be embedded into i_S. We first shift i_S leftwardly for one bit, and then set the last-significant bit (LSB) of the shifted result as w_2. For example, if $i_S = 6$, which can be expressed as 110 in binary form. After shifting 110 leftwardly for one bit, we have 1100. Now, if $w_2 = 1$, then $i_B = 1100 + w_2 = 1101$. Otherwise, $i_B = 1100 + w_2 = 1100$. Here we use i_B to denote the output index which contains both watermark bits. We summarise these steps as:

$$i_B = (i_S << 1) + w_2. \tag{9.5}$$

C. Index Assignment

After obtaining i_B from the above embedding procedure, the index assignment procedure introduced in Sect. 9.1.2 is employed now to produce two descriptions d_1 and d_2. Here, the index assignment matrix shown in Fig. 9.8 is referred since the length of i_B has been expanded. For instance, if $i_B = j_{25}$, then by referring to Fig. 9.8, we can easily find that $d_1 = 1000$ and $d_2 = 1001$.

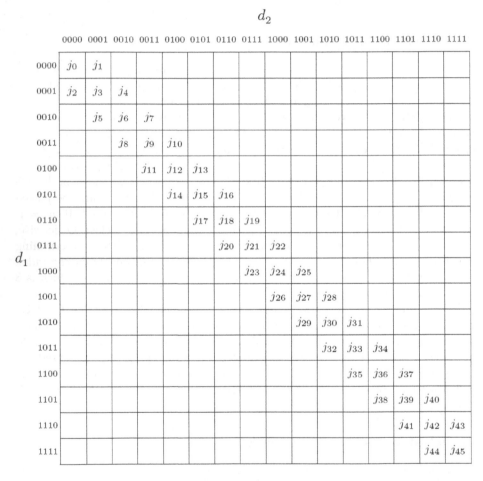

$$d_2$$

d_1	0000	0001	0010	0011	0100	0101	0110	0111	1000	1001	1010	1011	1100	1101	1110	1111
0000	j_0	j_1														
0001	j_2	j_3	j_4													
0010		j_5	j_6	j_7												
0011			j_8	j_9	j_{10}											
0100				j_{11}	j_{12}	j_{13}										
0101					j_{14}	j_{15}	j_{16}									
0110						j_{17}	j_{18}	j_{19}								
0111							j_{20}	j_{21}	j_{22}							
1000								j_{23}	j_{24}	j_{25}						
1001									j_{26}	j_{27}	j_{28}					
1010										j_{29}	j_{30}	j_{31}				
1011											j_{32}	j_{33}	j_{34}			
1100												j_{35}	j_{36}	j_{37}		
1101													j_{38}	j_{39}	j_{40}	
1110														j_{41}	j_{42}	j_{43}
1111															j_{44}	j_{45}

Fig. 9.8. An example of the combination of MDC and watermarking, an extension to Fig. 9.4(b)

9.2.2 Watermark Extraction

Corresponding to the watermark embedding algorithm, we describe the procedure, as depicted in Fig. 9.9, for extracting the two embedded watermarks. As an

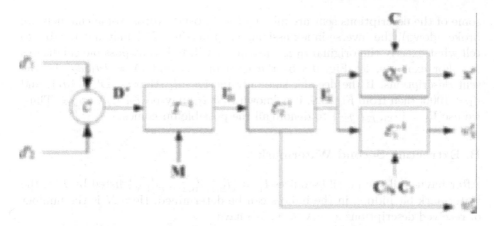

Fig. 9.9. The structure for extracting two watermark bits with two descriptions for transmission in MDC

inverse procedure to the embedding process, the secondly embedded watermark is extracted firstly, and the firstly embedded one is extracted then.

Step 1: Execute the collecting procedure $\mathcal{C}$ to collect all the descriptions received from the channels. Here due to some noise may occur during the transmission, some channels may break down and thus part of the descriptions sent may not be received. We use $\mathbf{D}'$ to denote the set containing all the descriptions received.

Step 2: Do the inverse index assignment procedure $\mathcal{I}^{-1}$ to determine the index which contains both watermark bits from $\mathbf{D}'$ by referring to the index assignment matrix $\mathbf{M}$. We use $\mathbf{I}'_B$ to denote the index set which contains all the possible recovered i_B.

Step 3: Apply the extracting procedure $\mathcal{E}_2^{-1}$ to determine what the watermark bit hidden in $\mathbf{I}'_B$ is. The determined result is denoted as w'_2, and the indics generated by discarding the watermark bit from $\mathbf{I}'_B$ are collected as $\mathbf{I}'_S$.

Step 4: Perform another extarcting procedure $\mathcal{E}_1^{-1}$ with subcodebooks $\mathbf{C}_0$ and $\mathbf{C}_1$ to determine what bit is carried by $\mathbf{I}'_S$. Details of this step are explained in the following section. We use w'_1 to denote the extracted watermark bit.

Step 5: Finally, execute the inverse VQ procedure $\mathcal{Q}_V^{-1}$ to reconstruct the output image, denoted as $\mathbf{x}'$, from $\mathbf{I}'_S$.

After all the blocks are reconstructed from the received descriptions by applying the above procedure one by one, we now have $\{\mathbf{x}'_0, \mathbf{x}'_1, ..., \mathbf{x}'_{N-1}\}$. Then, by piecing together those blocks, a reconstructed image $\mathbf{X}'$ can be formed.

A. Inverse Index Assignment

On the decoder side in Fig. 9.9, it first determines the possible outcomes $\mathbf{I}'_B$ from the received descriptions $\mathbf{D}'$. As we have addressed in Sect. 9.1.2.B, when some of the descriptions sent are missing (or, in other words, some channels are broken down), the inverse index assignment procedure $\mathcal{I}^{-1}$ may not be able to tell what exactly the original i_B is. Instead, it will tell what possible candicates are. For example, let Fig. 9.8 be the used matrix and $\mathbf{D} = \{d_1, d_2\}$ be the sent descriptions. If the decoder side only receives d_1, that is, $\mathbf{D}' = \{d_1\}$, and $d_1 = 1000$, then from Fig. 9.9, it is known that i_B may be j_{23}, j_{24}, or j_{25}. Thus, we use $\mathbf{I}'_B = \{j_{23}, j_{24}, j_{25}\}$ to denote all the possible outcomes.

B. Extracting Second Watermark

After having all the possible indices $\mathbf{I}'_B = \{i'_{B,1}, i'_{B,2}, ..., i'_{B,N}\}$ listed by $\mathcal{I}^{-1}$, the watermark bit hidden in the indices can be determined. Here N is the number of received descriptions and $N \leq K$. We have

$$w'_{2,p} = (i'_{B,p} \text{ MOD } 2), p = 1, 2, ..., N. \tag{9.6}$$

After having all the possible extracted bits $\{w'_{2,1}, w'_{2,2}, ..., w'_{2,N}\}$, the numbers of bit-0 and bit-1 are counted. Then, by employing the majority voting strategy, the embedded bit, denoted as w'_2, can be estimated. That is, if most of the extracted bits are 0, w'_2 is set as 0; if most of the extracted bits are 1, w'_2 is set as 1. In the case of N is even and the numbers of bit-0 and bit-1 are the same, we set w'_2 as 0 or 1 randomly.

Afterwards, each $i'_B \in \mathbf{I}'_B$ is shifted to the right by one bit to smooth away the effects from watermark embedding. We have:

$$i'_{S,p} = (i'_{B,p} >> 1), p = 1, 2, ..., N. \tag{9.7}$$

The recovered indices $\{i'_{S,1}, i'_{S,2}, ..., i'_{S,N}\}$ are collected as $\mathbf{I}'_S$ and output to the next step.

C. Extracting First Watermark

After obtaining the possible indices, we are able to determine what watermark bits are carried by them. Here for either index $i'_{S,p} \in \mathbf{I}'_S$, where $1 \leq p \leq N$, the codeword whose index is $i'_{S,p}$ can be picked from $\mathbf{C}$. We examine whether this codeword belongs to the sub-codebook $\mathbf{C}_0$ or $\mathbf{C}_1$, assume it is $\mathbf{C}_q$, where $q \in [0, 1]$, and set the watermark bit $w'_{1,p}$ as q. For example, if the codeword belongs to $\mathbf{C}_0$, then $w'_{1,p} = 0$, otherwise, $w'_{1,p} = 1$. By repeating the above step for N times, the watermark bits carried by $\{i'_{S,1}, i'_{S,2}, ..., i'_{S,N}\}$ can be determined respectively. We now have N watermark bits $\{w'_{1,1}, w'_{1,1}, ..., w'_{1,N}\}$. Then, the same as Sect. 9.2.2.B, we count the numbers of bit-0 and bit-1 of these watermark bits, and set $w'_1 = 0$ if most of the extracted bits are 0, or set $w'_1 = 1$ if most of the extracted bits are 1. In the case of N is even and the numbers of bit-0 and bit-1 are the same, we set w'_1 as 0 or 1 randomly.

D. Reconstructing Output Image

As the decoder side may receive more than one index for an input image block, this means that more than one image block will be reconstructed and we may have several versions of reconstructed image blocks. For example, assume $\mathbf{I}'_S$ is a set containing M possible indices $\{i'_{S,1}, i'_{S,2}, ..., i'_{S,M}\}$. Then, after performing the inverse quantisation step, we have N different versions of reconstructed image blocks $\{\mathbf{x}'_1, \mathbf{x}'_2, ..., \mathbf{x}'_M\}$. In some cases, having the unique reconstructed image may be important. Thus, Eq. (9.4) and the step presented in Sect. 9.1.3.B can be referred to.

9.3   Codebook Partition

The main problem for the first watermark embedding algorithm in Sect. 9.2.1.A is how to split the codebook $\mathbf{C}$ into two sub-codebooks $\mathbf{C}_0$ and $\mathbf{C}_1$. Also, the result for splitting $\mathbf{C}$ will not only influence the watermark imperceptibility and the robustness of the first watermark, but it will also effect the robustness of the second watermark. All the problems can be optimized with tabu search [49] by offering the reasonable fitness function. With the fundamentals described in Sect. 3.5, both the imperceptibility of the watermarked image, represented by PSNR (see Eq. (2.4)), and the robustness of the extracted watermarks represented by BCR (see Eq. (2.7)), can be considered for optimization. The fitness function suggested in [66] is:

$$ f_i = \text{PSNR}_i + \lambda_1 \times \text{BCR}_{1,i} + \lambda_2 \times \text{BCR}_{2,i}, \tag{9.8} $$

where f_i denotes the fitness score in the i-th iteration, PSNR_i is the PSNR value in the i-th iteration, and $\text{BCR}_{1,i}$ and $\text{BCR}_{2,i}$ denote the BCR values in the i-th iteration. Here because the PSNR values are generally dozens of times larger than the BCR values, λ_1 and λ_2 are introduced to represent the weighting factors to balance the effects from PSNR and BCR. The objective is to maximize f_i in our system.

From the preliminaries given in Sect. 6.4, we expect that by fixing the watermark capacity, both the watermark imperceptibility and watermark robustness can be improved after the optimization.

9.4   Simulation Results

In the simulations presented in [66], the image of LENA (see Fig. 1.2) with size of 512×512 in gray level was used as the cover image $\mathbf{X}$. The binary images shown in Fig. 9.10(a) were used as the watermarks. Here the image of Rose with size 128×128 was used as the first watermark $\mathbf{W}_1$, and the other one, which also has the size of 128×128, was used as the second watermark $\mathbf{W}_2$. Two codebooks with sizes $L = 512$ and $L = 1024$ were used, and the indices therein were represented by 9-bit and 10-bit strings, respectively.

The cover image was decomposed into 128×128 (which exactly meets the number of watermark bits) non-overlapping blocks, thus the size of each block is $\frac{512 \times 512}{128 \times 128} = 4 \times 4$. To split the codebooks into two sub-codebooks, the mentioned codebook partition procedure using tabu search (see Sect. 3.5) was applied. Here the settings used in the training procedure are: 20 candidate solutions trained for each iteration; the tabu list length T_S is set to 10; the weighting factors are set to $\lambda_1 = \lambda_2 = 10$; the aspiration value is set to 40, with $\text{PSNR}_i \geq 26$, $\text{BCR}_{1,i} \geq 0.7$, and $\text{BCR}_{2,i} \geq 0.7$ in Eq. (9.8); the number of total training iterations is set to 100. Watermark embedding and extraction were performed in every training iteration to obtain the updated PSNR_{i+1}, $\text{BCR}_{1,i+1}$, and $\text{BCR}_{2,i+1}$ in the next iteration.

To evaluate the performace of the watermarking system, PSNR and BCR were employed to describe the quality of the wateramrked result and the robustness of the system, respectively. Generally specaking, the higher values the PSNR and BCR are, the better performace the system possesses.

As shown in Fig. 9.1, only the descriptions of the watermarked VQ indices are transmitted over the noisy channels. Therefore, the general attacking schemes, such as low-pass filtering, or those employed in the Stirmark benchmark [63], are not suitable to be applied. Here, we only focus on the situations for transmitting the descriptions over mutually independent channels.

Simulation results with different channel erasure probabilities are presented in Table 9.1 while the codebook with length $L = 512$ was used. The watermarks extracted are shown in Fig. 9.10 and Fig. 9.11. Table 9.2 lists the results while the codebook with length $L = 1024$ was used. Figures 9.12 and 9.13 show the watermarks extracted. In the tables and figures presented, p_1 and p_2 denote the erasure probabilities with Channel 1 and Channel 2, respectively.

As Fig. 9.10(a) and Fig. 9.12(a) shown, while the descriptions were sent via the error-free channels (that is $p_1 = p_2 = 0.0$), the watermarks extracted are identical with the embedded ones. In Fig. 9.10(b)-(d) and Fig. 9.12(b)-(d), they represent the results for transmitting over the lightly to heavily erased channels. The BCR values are high and the extracted watermarks are recognizable even

Table 9.1. Watermarked image quality under different channel erasure probabilities for the codebook length $L = 512$

Channel erasure probabilities		PSNR (dB)	BCR (%)	
p_1	p_2		$\mathbf{W}'_1$	$\mathbf{W}'_2$
0.0	0.0	30.74	100	100
0.1	0.1	28.15	94.12	93.60
0.25	0.25	24.39	87.50	85.64
0.5	0.25	19.88	74.30	72.08
0.0	1.0	26.19	81.03	68.23
1.0	0.0	26.13	67.00	69.71

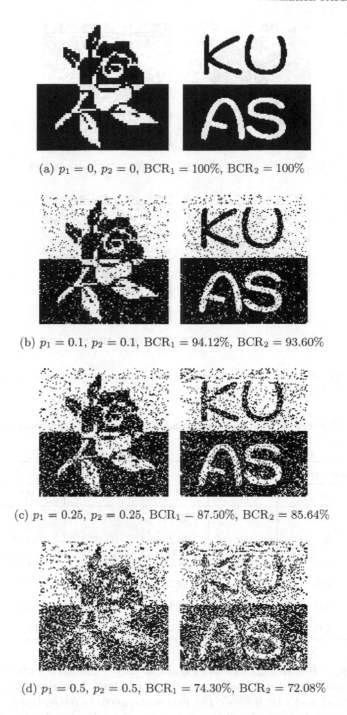

(a) $p_1 = 0$, $p_2 = 0$, BCR$_1$ = 100%, BCR$_2$ = 100%

(b) $p_1 = 0.1$, $p_2 = 0.1$, BCR$_1$ = 94.12%, BCR$_2$ = 93.60%

(c) $p_1 = 0.25$, $p_2 = 0.25$, BCR$_1$ − 87.50%, BCR$_2$ = 85.64%

(d) $p_1 = 0.5$, $p_2 = 0.5$, BCR$_1$ = 74.30%, BCR$_2$ = 72.08%

Fig. 9.10. The two extracted watermarks under different channel erasure probabilities, with codebook size $L = 512$

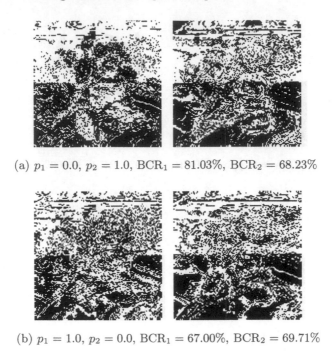

(a) $p_1 = 0.0$, $p_2 = 1.0$, $\text{BCR}_1 = 81.03\%$, $\text{BCR}_2 = 68.23\%$

(b) $p_1 = 1.0$, $p_2 = 0.0$, $\text{BCR}_1 = 67.00\%$, $\text{BCR}_2 = 69.71\%$

Fig. 9.11. The two extracted watermarks under the case when one channel is total breakdown, with codebook size $L = 512$.

Table 9.2. Watermarked image quality under different channel erasure probabilities for the codebook length $L = 1024$

Channel erasure probabilities		PSNR (dB)	BCR (%)	
p_1	p_2		$\mathbf{W}_1'$	$\mathbf{W}_2'$
0.0	0.0	32.74	100	100
0.1	0.1	28.35	94.27	93.44
0.25	0.25	24.27	86.52	84.78
0.5	0.25	19.78	73.28	71.34
0.0	1.0	25.10	76.61	70.78
1.0	0.0	25.04	68.56	70.26

with $p_1 = p_2 = 0.5$. In Fig. 9.11 and Fig. 9.13, they demonstrate when one of the channels experiences total breakdown. When Channel 2 breaks down, as shown in Fig. 9.11(a) and Fig. 9.13(a), the first watermark is recognizable, while the second watermark cannot be distinguished. On the other hand, when Channel 1 breaks down, as shown in Fig. 9.11(b) and Fig. 9.13(b), we obtain similar results to those in either Fig. 9.11(a) or Fig. 9.13(a).

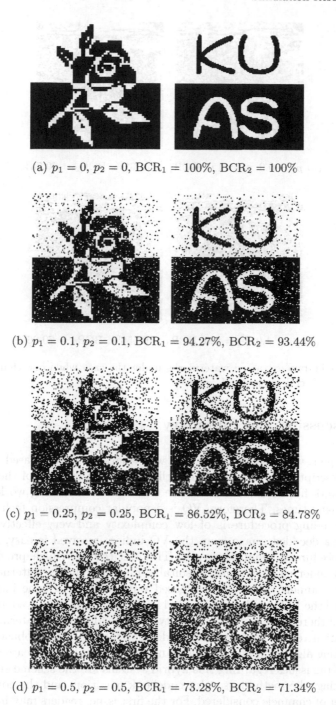

(a) $p_1 = 0$, $p_2 = 0$, $\text{BCR}_1 = 100\%$, $\text{BCR}_2 = 100\%$

(b) $p_1 = 0.1$, $p_2 = 0.1$, $\text{BCR}_1 = 94.27\%$, $\text{BCR}_2 = 93.44\%$

(c) $p_1 = 0.25$, $p_2 = 0.25$, $\text{BCR}_1 = 86.52\%$, $\text{BCR}_2 = 84.78\%$

(d) $p_1 = 0.5$, $p_2 = 0.5$, $\text{BCR}_1 = 73.28\%$, $\text{BCR}_2 = 71.34\%$

Fig. 9.12. The two extracted watermarks under different channel erasure probabilities, with codebook size $L = 1024$

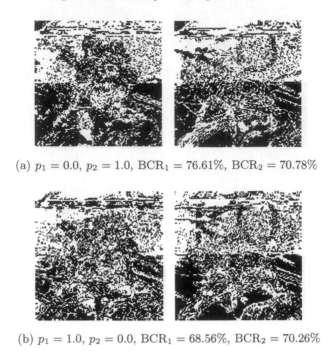

(a) $p_1 = 0.0$, $p_2 = 1.0$, $BCR_1 = 76.61\%$, $BCR_2 = 70.78\%$

(b) $p_1 = 1.0$, $p_2 = 0.0$, $BCR_1 = 68.56\%$, $BCR_2 = 70.26\%$

Fig. 9.13. The two extracted watermarks under the case when one channel is total breakdown, with codebook size $L = 1024$

9.5   Discussion and Summary

In this chapter, we introduced a hybrid watermarking system based on VQ and multiple description coding. As it is well known that in most of the cases, the encoding work has to be executed once only but the decoding work has to be done for many times. Therefore, VQ is a popular compression technique since the VQ decoding procedure is of low complexity and very effective. That is, generating a decoded image from the VQ indices received is easy, as a table-lookup procedure is required merely. The MD coding, which provides better robustness under channel noise, is suitable for transmitting watermark signals over noisy channels. Also, the VQ-based watermarking scheme introduced in Chap. 6 and the LSB method illustrated in Chap. 4 were employed and utilized. The system therefore posseses the ability of embedding dual watermarks. From the simulation results presented, they demonstrate both the robustness of the watermarking algorithm and the resistance under channel noise are good.

Some of the issues regarding this hybrid systems are the scheme employed for watermarking, the method used for codebook partition, and the complexity of the number of channels considered. For the first issue, readers may have already seen that the watermarking procedure of the hybrid system can be replaced by either suitable scheme. For example, the schemes introduced in Chap. 6 can be employed. Designers who want to implement this system can alter this part of the system according to their need. For the second issue, although the tabu search was employed in the simulation to split the codebook, designers still can employ other method to play the act. Here readers can refer to Sect. 6.4, which employs GAs, for more details. As to the questions about what complexity and performance the system will have while the number (K) of more channels are considered, the authors who proposed this hybrid system did not address furthermore. Here we leave them to the readers and encourage the readers to investigate them.

To sum up, the hybrid watermarking system not only provides good performance under noisy channels, but also posseses the advantages such as easy to implement, low complexity in decoding, and dual-watermark capacity. This indicates the watermarking system introduced is not only innovative for researching, but also suitable for practical implementations.

Chapter 10
Fake Watermark Embedding Scheme Based on Multi-Stage VQ

A multiple fake-watermark embedding system based on multi-stage vector quantisation (MSVQ) for security enhancement is given in this chapter. This system first creates a non-recognizable reference watermark by referring to the genuine watermark and a number of fake watermarks. It then performs the MSVQ to encode the cover image. In each stage of the MSVQ system, a polarity stream is established according to the VQ indices obtained. The exclusive-OR (XOR) operation is then applied to the polarity and the corresponding watermark, which may be a fake watermark or the reference watermark, so that a user key can be created. To reveal the genuine watermark, the MSVQ is first performed on the watermarked image. In each stage of the MSVQ system, a watermark can be extracted using the key generated previously. By stacking all the extracted watermarks together, the original watermark can be recovered.

Due to the use of visible fake watermarks, the security of the watermarking system is improved. Moveover, by assigning the keys generated to the related users, the introduced system allows multi-users to share the hidden information together.

10.1 Introduction

For a watermarking system, the security is sometimes of greater concern than other benchmarks such as robustness or capacity. Among the methods proposed for increasing the security, using fake watermarks is one of the straightforward solutions since the unauthorized people who try to obtain the hidden information may be convinced easily by the content of the extracted fake watermarks.

In this chapter, designing a watermarking system using fake watermarks is the main focus. The concepts of MSVQ (multi-stage vector quantisation) [24] and the VQ-based watermarking schemes proposed in [39] [40] are introduced first in Sect. 10.1 as the background knowledge. After that, the method for coding the fake watermarks and the watermarking scheme based on MSVQ and [39] [40] are described in Sect. 10.2. Simulation results presented in Sect. 10.3 will demonstrate the performance of the introduced watermarking method.

F.-H. Wang, J.-S. Pan, and L.C. Jain: Innovations in Dig. Watermark. Tech., SCI 232, pp. 151–162.
springerlink.com   © Springer-Verlag Berlin Heidelberg 2009

10.1.1 Multi-Stage Vector Quantisation (MSVQ)

Compared to the traditional VQ system (Sect. 6.2), a MSVQ system [24] possesses the advantages of shorter codeword-search time and smaller storage space for codebooks. The main concept of the MSVQ system is to divide the traditional VQ system into several subsystems (or stages). Each of the subsystems does the same duty. Due to this, the size of the codebook used in each stage can be reduced. The search time for obtaining the nearest codewords is therefore shortened then.

An illustration of a n-stage MSVQ system is shown in Fig. 10.1, where $\mathbf{X}$ is the input image of the MSVQ system; $\mathbf{E}_i$, $\mathbf{C}_i$, $\mathbf{X}_i$, and $\mathbf{I}_i$ are the input image, the codebook used, the output image, and the index set of the i-th stage, respectively. The definition of $\mathbf{E}_i$ is:

$$\mathbf{E}_i = \begin{cases} \mathbf{X}, & \text{if } i = 1 ; \\ \mathbf{E}_{i-1} - \mathbf{X}_{i-1}, & \text{if } i > 1. \end{cases} \tag{10.1}$$

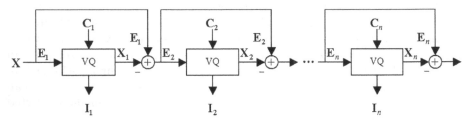

(a) The encoding procedure

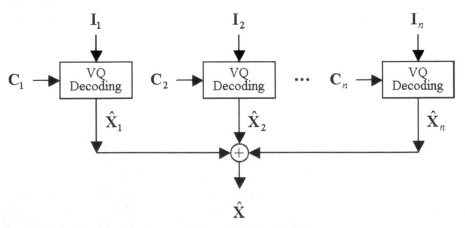

(b) The decoding procedure

Fig. 10.1. The block diagrams of the MSVQ system

The encoding procedure for each stage is the same as the traditional VQ encoding procedure. The input image $\mathbf{E}_i$ of the i-th stage is first decomposed into many non-overlapping vectors with size d pixels. Then, a nearest-codeword-search procedure (Sect. 6.2.2) is performed to find the codeword with the minimum distortion from $\mathbf{C}_i$ for each vector. By assembling all the codewords obtained, a reconstructed image $\mathbf{X}_i$ can be formed. Equation (10.1) is then used to establish the input image $\mathbf{E}_{i+1}$ for the next stage. The indices of the codewords obtained are collected as the set $\mathbf{I}_i$ and it is then delivered to the receiver.

On the receiver side, when the index set $\mathbf{I}_i$ from the i-th stage is received, the normal VQ decoding procedure introduced in Sect. 6.2.3 is done to reconstruct a decoded image $\hat{\mathbf{X}}_i$. After all the reconstructed images $\{\hat{\mathbf{X}}_1, \hat{\mathbf{X}}_2, ..., \hat{\mathbf{X}}_n\}$ have been available, the final reconstructed image $\hat{\mathbf{X}}$ can be established by summing up $\{\hat{\mathbf{X}}_1, \hat{\mathbf{X}}_2, ..., \hat{\mathbf{X}}_n\}$, as shown in Fig. 10.1(b). The example of applying the MSVQ procedure to the image of LENA is given in Fig. 10.2.

10.1.2   MSVQ-Based Watermarking Scheme

In this section, the VQ-based watermarking scheme proposed in [39] is introduced. Figure 10.3 displays its block diagrams.

Let $\mathbf{X}$ be a cover image with size $M \times H$ pixels and $\mathbf{W}$ be a binary-valued watermark with size $M_W \times H_W$ pixels. In order to survive the picture-cropping attacks, a pseudo-random-number traversing method [37] is suggested. This technique is performed to disperse the spatial-domain relationships of $\mathbf{W}$. That is:

$$\mathbf{W}_P = \text{Permute}(\mathbf{W}, seed). \tag{10.2}$$

where $seed$ is a user-selected seed for the pseudo-random number generator and $\mathbf{W}_P$ is the permuted result of $\mathbf{W}$.

The cover image $\mathbf{X}$ is decomposed into a number of vectors with size $m \times h$ pixels, and the VQ encoding procedure (Sect. 6.2.2) is then performed to obtain the indices of the nearest codewords for all the vectors. Here let the vector at the position (r, s) of $\mathbf{X}$ be $\mathbf{x}(r, s)$, where $0 \le r < \frac{M}{m}$ and $0 \le s < \frac{H}{h}$. The index $y(r, s)$ of the codeword obtained for $\mathbf{x}(r, s)$ can be expressed with

$$y(r, s) = \text{VQ}(\mathbf{x}(r, s)) \in [0, L), \tag{10.3}$$

where L is the size of the codebook used. The indices of all the nearest codewords are then collected as the index set $\mathbf{Y}$:

$$\mathbf{Y} = \bigcup_{i=0}^{\frac{M}{m}-1} \bigcup_{j=0}^{\frac{H}{h}-1} \{y(i, j)\} = \text{VQ}(\mathbf{X}). \tag{10.4}$$

To embed the watermark $\mathbf{W}$ into the cover image $\mathbf{X}$, we need to introduce some relationships to transform the VQ indices into binary formats for further embedding. Hence, we bring up the polarities $\mathbf{P}$ of the VQ indices to embed the watermark. For natural images, the VQ indices among the neighboring blocks tend to be similar. We can make use of this characteristic to generate $\mathbf{P}$ by

(a) The original image

(b) The MSVQ coded image

Fig. 10.2. The example of applying MSVQ ($n = 3$) to the image of LENA (PSNR=30.19 dB)

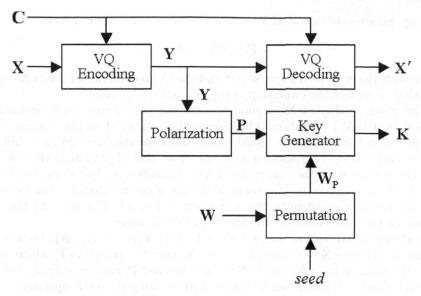

(a) The embedding procedure

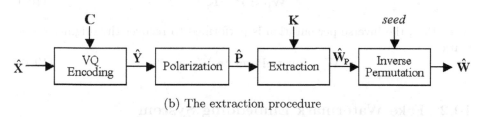

(b) The extraction procedure

Fig. 10.3. The block diagrams of Huang-Wang-Pan's watermarking scheme

calculating the mean values for each indices. That is, for the index $y(r,s)$ at the position (r,s) of $\mathbf{Y}$, where $0 \leq r < \frac{M}{m}$ and $0 \leq s < \frac{H}{h}$, the mean value of $y(r,s)$ and its surrounding indices is:

$$\mu(r,s) = \frac{1}{9} \sum_{i=r-1}^{r+1} \sum_{j=s-1}^{s+1} y(i,j) . \qquad (10.5)$$

Based on the mean value, the polarity of $y(r,s)$ can be decided by:

$$p(r,s) = \begin{cases} 1, & \text{if } y(r,s) \geq \mu(r,s) ; \\ 0, & \text{otherwise.} \end{cases} \qquad (10.6)$$

The polarities established are then collected to form the polarity set $\mathbf{P}$:

$$\mathbf{P} = \bigcup_{i=0}^{\frac{M}{m}-1} \bigcup_{j=0}^{\frac{H}{h}-1} \{p(i,j)\} . \qquad (10.7)$$

Finally, we are able to embed $\mathbf{W}_P$ with $\mathbf{P}$ by using the XOR operator:

$$\mathbf{K} = \mathbf{W}_P \oplus \mathbf{P} \, , \tag{10.8}$$

where $\mathbf{K}$ is the generated secret key. It and the VQ reconstructed image $\mathbf{X}'$ work together to protect the ownership of the original cover image.

The visual quality of $\mathbf{X}'$ is good because the only distortion introduced is caused by the VQ operation. The information conveyed in the watermarked image will not influence the visual quality because the information is hidden in the secret key. From another point of view, this algorithm is efficient for implementation with the conventional VQ compression algorithms. Once the codeword for each vector is chosen, it is then able to determine the polarity of each vector; consequently, the secret key is formed. The key and the VQ reconstructed image are then transmitted to the receiver.

In the watermark extraction procedure, first the VQ operation is performed on the received image $\hat{\mathbf{X}}$ to obtain the indices $\hat{\mathbf{Y}}$, then the polarities $\hat{\mathbf{P}}$ is estimated from the mean value calculated. With the obtained $\hat{\mathbf{P}}$ and the existing key $\mathbf{K}$, the embedded watermark can be determined by using the XOR operator:

$$\hat{\mathbf{W}}_P = \hat{\mathbf{P}} \oplus \mathbf{K} \, . \tag{10.9}$$

After that, the inverse permutation is performed to recover the original watermark:

$$\hat{\mathbf{W}} = \text{InversePermute}(\hat{\mathbf{W}}_P, seed) \, . \tag{10.10}$$

10.2   Fake Watermark Embedding System

In this section, the procedures for encrypting and decrypting the fake watermarks with the real watermark are described first. The proposed watermarking scheme based on MSVQ and [39] is then illustrated.

10.2.1   Encryption and Decryption for Watermarks

Let $\mathbf{W}$ be the genuine watermark, $\mathbf{W}_F = \{\mathbf{W}_{F_1}, \mathbf{W}_{F_2}, ..., \mathbf{W}_{F_n}\}$ be a set of n recognizable fake watermarks, and $\mathbf{W}_R$ be an unrecognizable reference watermark. Our goal here is to mix up the real watermark and fake watermarks to generate an unrecognizable reference watermark according to a user-selected seed for future use, that is:

$$\mathbf{W}_R = \text{Encrypt}(\mathbf{W}, \mathbf{W}_F, seed) \, . \tag{10.11}$$

The encryption process is divided into two sub-processes. First, Eq. (10.12) is applied to generate a temporary watermark $\mathbf{W}_T$.

$$\mathbf{W}_T = \mathbf{W} \oplus \mathbf{W}_{F_1} \oplus \mathbf{W}_{F_2} \oplus \cdots \oplus \mathbf{W}_{F_n} \, . \tag{10.12}$$

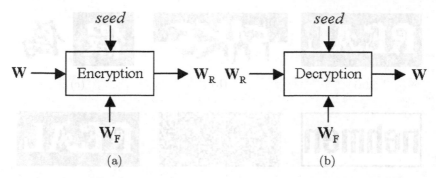

Fig. 10.4. The block diagrams for encrypting and decrypting the fake watermarks with the real watermark

A pseudo-random-number traversing method is then done to disperse the spatial-domain relationships of $\mathbf{W}_T$. By referring to a user-selected seed, we have:

$$\mathbf{W}_R = \text{Permute}(\mathbf{W}_T, seed) \,, \tag{10.13}$$

where $\mathbf{W}_R$ is the reference watermark, which is the permuted version of $\mathbf{W}_T$.

The decryption procedure for obtaining the hidden real watermark is simple. First the inverse permutation is performed to generate a temporary watermark $\mathbf{W}_T$ using the same seed used in the encryption procedure:

$$\mathbf{W}_T = \text{InversePermute}(\mathbf{W}_R, seed) \,. \tag{10.14}$$

The XOR operation is then performed on $\mathbf{W}_T$ and all the fake watermarks to recover the original watermark $\mathbf{W}$:

$$\mathbf{W} = \mathbf{W}_T \oplus \mathbf{W}_{F_1} \oplus \mathbf{W}_{F_2} \oplus \cdots \oplus \mathbf{W}_{F_n} \,. \tag{10.15}$$

We use Eq. (10.16) to replace Eqs. (10.14) and (10.15) for expressing the decryption procedure.

$$\mathbf{W} = \text{Decrypt}(\mathbf{W}_R, \mathbf{W}_F, seed) \,. \tag{10.16}$$

An example of the introduced method using three fake watermarks is shown in Fig. 10.5.

10.2.2 Embedding and Extraction Procedures

The goal of the watermarking scheme introduced in this section is to hide the real watermark using some fake watermarks to make the security of the system better and safer. The encryption procedure is first performed to create a reference watermark $\mathbf{W}_R$ from the real watermark $\mathbf{W}$ and the $(n-1)$ fake watermarks $\{\mathbf{W}_{F_1}, \mathbf{W}_{F_2}, ..., \mathbf{W}_{F_{n-1}}\}$. Afterwards, the n-stage MSVQ operation is performed on the cover image $\mathbf{X}$. With the indices obtained in each stage, Eqs. (10.5)~(10.7)

Fig. 10.5. An example of the introduced encryption and decryption schemes. (a) The real watermark. (b)~(d) Three fake watermarks. (e) The generated reference watermark. (f) The recovered watermark.

are used to establish the polarity stream for embedding. The reference watermark $\mathbf{W_R}$ is assigned to be embedded in the first stage, the first fake watermark is assigned to be embedded in the second stage, and so on. Finally, each stage generates a user key. For the purpose of sharing secrets within multi-users, these keys can be assigned to the related users. Otherwise, they are collected as $\mathbf{K}$ and is assigned to the single user.

In the extraction procedure, the n-stage MSVQ operation is performed first to obtain the indices. With the indices, Eqs. (10.5)~(10.7) are used again to generate the polarity stream. With the polarity stream and the secret keys that generated in each stage of the embedding procedure, we can extract the embedded watermarks stage by stage. Finally, the decryption process is carried out upon all the watermarks extracted to recover the real one. The block diagrams for illustrating the embedding and the extraction procedures are shown in Fig. 10.6.

10.3   Simulation Results

In our simulation, the image of LENA shown in Fig. 10.2(a) was used as the host image. Its size is 512×512 pixels in gray-level. The images shown in Figs. 10.7(a), (b), and (c) were used as the genuine watermark and the two fake watermarks, respectively. All of the watermarks are binary images with size 128×128 pixels. The number of stages in this MSVQ system is $n = 3$. The codebook size in each stage is $L = 8$.

We first encrypted the real watermark with the fake watermarks to generate a reference watermark (Fig. 10.7(d)). This and the other two fake watermarks were then embedded in the corresponding stages of the MSVQ system. The PSNR (Eq. (2.4)) value between the original cover image and the output image (Fig. 10.2(b)) is 30.19 dB.

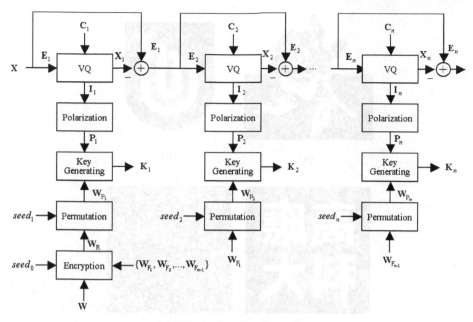

(a) The embedding procedure

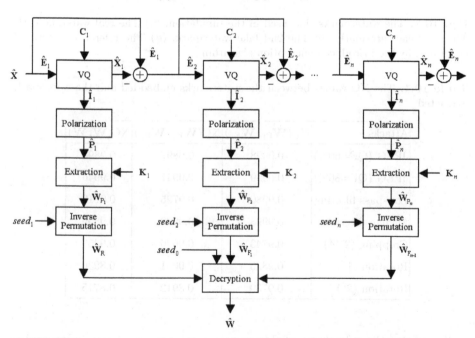

(b) The extraction procedure

Fig. 10.6. The block diagrams of the MSVQ-based watermarking system

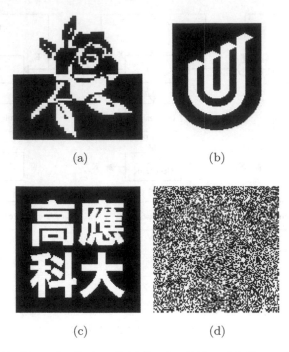

(a)   (b)

(c)   (d)

Fig. 10.7. The watermarks that used in the simulation. (a) The real watermark. (b) The 1st fake watermark. (c) The 2nd fake watermark. (d) The reference watermark generated by the introduced encryption algorithm.

Table 10.1. The NC values between the watermarks embedded and the watermarks extracted

Attacks	$NC(\mathbf{W}_{F_1}, \hat{\mathbf{W}}_{F_1})$	$NC(\mathbf{W}_{F_2}, \hat{\mathbf{W}}_{F_2})$	$NC(\mathbf{W}, \hat{\mathbf{W}})$
JPEG (QF=60%)	0.9998	0.9897	0.9894
JPEG (QF=80%)	0.9999	0.9947	0.9946
Low-pass filtering	0.9980	0.9725	0.9712
Median filtering	0.9998	0.9900	0.9899
Cropping (25%)	0.9743	0.9626	0.9453
Rotation (1°)	0.9825	0.9051	0.8930
Rotation (2°)	0.9739	0.8912	0.8715

For testing the robustness of this system, some attack schemes were applied to attack the watermarked image. These schemes are the JPEG compression with different quality factors (QF), low-pass filtering, median filtering, cropping, and rotation. Figure 10.8 shows the genuine watermarks recovered from the watermarks which are extracted from the attacked images. Table 10.1 lists the

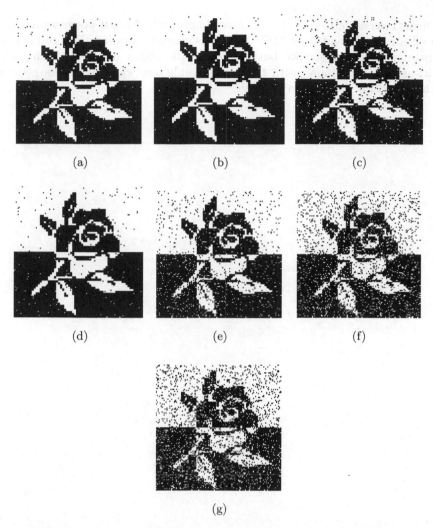

Fig. 10.8. The recovered watermarks after decrypting all the watermarks extracted from the attacked images: (a) JPEG compression with a QF=60%, (b) JPEG compression with a QF=80%, (c) low-pass filtering with a window size =3, (d) median filtering with a window size =3, (e) cropping 25% in the lower-left quarter, (f) rotation with 1°, and (g) rotation with 2°.

NC (Eq. (2.3)) values between the embedded watermarks and the extracted watermarks under the attacks mentioned.

10.4 Summary

Watermarking using fake watermarks based on MSVQ has been presented in this chapter. This system is easy to implement and robust to some common attacks, especially the most popular JPEG compression. The simulation results have confirmed the usefulness of the introduced approach. Also, it possesses not only the advantages provided by MSVQ, such as shorter encoding time and smaller storage space for codebooks, but it also has the ability of secret sharing with multi-users. Furthermore, with the use of visible fake watermarks, the security of this system becomes even stronger.

Chapter 11
Watermarking with Visual Cryptography and Gain-Shape VQ

Employing the concept of visual cryptography to design a watermarking scheme is introduced in this chapter. A modified visual cryptography is applied to split the genuine watermark into two shadow watermarks. The gain-shape vector quantisation (GSVQ) procedure is then performed to encode the cover image. Afterwards, the shadow watermarks and the VQ indices obtained are processed to generate two user-keys.

The watermarking system presented possesses a number of advantages, including easy implementation (the structure of the GSVQ system is not altered) and stronger security. Experimental results will show its performance.

11.1 Introduction

For hiding information within still images, visual cryptography (or visual secret sharing scheme, VSS scheme) [60] [75] is another well-known and popular method. A traditional VSS scheme splits a secret image into several meaningless images, called *shares*. From either of the shares generated, one cannot obtain any useful information relating to the original secret image. The only method which can reveal the secret image is to stack the considered number of shares together. Afterwards, the content of the secret image can be observed using bare human eyes directly.

As stated in Sect. 2.1.1, encoding and/or encrypting the original watermark before embedding it into the cover image helps improve the security. Therefore, the concept of applying visual cryptography in watermarking systems has been considered. Hou and Chen [31] proposed a watermarking scheme with a modified visual cryptography. They split the original binary watermark into two unrecognizable gray-valued shares. The first share is used as the input watermark and the second share is regarded as a secret key used for verification. To reveal the original watermark from the watermarked image, the secret share is merely stacked upon the watermarked image. If the original watermark can be recognized from the stacked result, the ownership of it can be claimed. Wang et al. [105] proposed a method to process repeating watermarks into two shares.

F.-H. Wang, J.-S. Pan, and L.C. Jain: Innovations in Dig. Watermark. Tech., SCI 232, pp. 163–172.
springerlink.com   © Springer-Verlag Berlin Heidelberg 2009

Similar to [31], they also used one share for embedding and used another share for verification. Other watermarking schemes using visual cryptography can be found in [18] [67]. In summary, these watermarking schemes apply visual cryptography to split the original watermark into two shares, and embed one of them into the cover image. The other unused share is regarded as a user key to verify the ownership of the watermarked image.

In this chapter, the background knowledge of visual cryptography and GSVQ [24] are given first in Sects. 11.1.1 and 11.1.2 respectively. The watermarking scheme based on VSS and GSVQ is then introduced in Sect. 11.2. This scheme splits the original watermark into two meaningless shares and applies the GSVQ procedure to encode the cover image. The indices and shares obtained are then encoded using the method introduced in Sect. 10.1.2 to generate two user-keys. Experimental results given in Sect. 11.3 will show its performance.

11.1.1 Visual Cryptography

Visual cryptography (or visual secret sharing scheme, VSS scheme) [60] [75] is an effective scheme for hiding a secret within a number of still images. A traditional VSS scheme randomly splits the secret image into n meaningless images, named *shares*, so that one cannot obtain any useful information regarding to the secret image from either of the shares. The hidden secret can be revealed if and only if the considered number m, where $m \leq n$, of shares are presented. By stacking these m shares together, the hidden secret can then be revealed without any calculation. To use visual cryptography for the purpose of secret sharing, the generated shares are assigned to the related users merely. The secret image therefore is shared by n users. A simple VSS scheme is described below.

Let $\mathbf{W}$ denote a binary image of size $M \times N$ pixels and $\mathbf{S}_1$ and $\mathbf{S}_2$ denote the share images of size $2M \times 2N$ pixels, which are split from $\mathbf{W}$. For each pixel of $\mathbf{W}$, according to its value (black or white), two blocks of size 2×2 pixels are selected randomly from Table 11.1. That is, if the input pixel is white, then a block pair is selected randomly from numbers 1 to 6 of Table 11.1. Otherwise, a block pair from numbers 7 to 12 of Table 11.1 is selected randomly. Afterwards, the blocks in the selected pair are assigned to $\mathbf{S}_1$ and $\mathbf{S}_2$ respectively. By repeating the above process to deal with all the pixels in $\mathbf{W}$, the blocks collected in $\mathbf{S}_1$ and

Table 11.1. The block pairs for the (2,2)-VSS scheme

Original pixel	White						Black					
Number	1	2	3	4	5	6	7	8	9	10	11	12
Block 1												
Block 2												
Stacked result												

S_2 are pieced together respectively to form two share images. These two share images are then printed on transparency slides or films.

To reveal the original image from the shares, it requires no computation. The transparency slides are stacked together merely. The secret image can then be observed by bare human eyes directly. An example of applying the splitting procedure to a given binary image is displayed in Fig. 11.1.

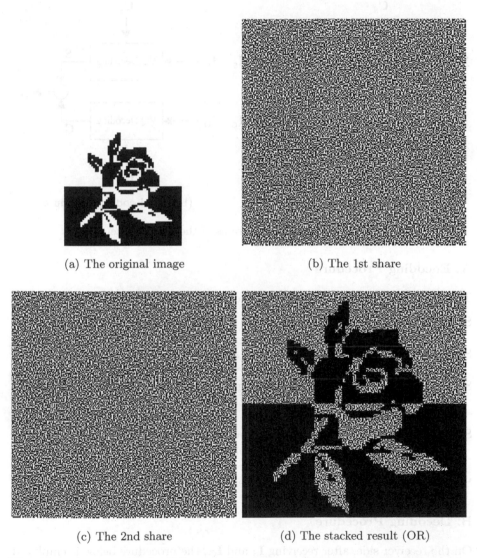

(a) The original image

(b) The 1st share

(c) The 2nd share

(d) The stacked result (OR)

Fig. 11.1. An example of splitting one binary image into two meaningless shares by using the (2,2)-VSS scheme

11.1.2 Gain-Shape Vector Quantisation (GSVQ)

The gain-shape VQ (or, shape-gain VQ) system [24] possesses a number of advantages, including a faster encoding time and a smaller codebook storage space, comparing with the traditional VQ system. Figure 11.2 shows the structure of a GSVQ system.

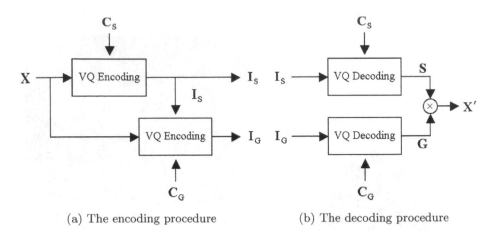

(a) The encoding procedure   (b) The decoding procedure

Fig. 11.2. The block diagrams of the GSVQ system

A. Encoding Procedure

The steps below illustrate the coding procedure of a GSVQ system:

Step 1: Decompose the input image $\mathbf{X}$ into T non-overlapping vectors $\{\mathbf{x}_1, \mathbf{x}_1, ..., \mathbf{x}_T\}$ of size d pixels.

Step 2: Perform the nearest codeword search (Sect. 6.2.2) to obtain a nearest shape codeword from the shape codebook $\mathbf{C}_S$ for each vector.

Step 3: According to the corresponding shape codeword obtained, perform the nearest codeword search again to obtain a nearest gain value from the gain codebook $\mathbf{C}_G$ for each vector.

Step 4: Collect the indices of all the shape codewords obtained in Step 2 as the shape index set $\mathbf{I}_S$ and the indices of all the gain values obtained in Step 3 as the gain index set $\mathbf{I}_G$.

Step 5: Transmit $\mathbf{I}_S$ and $\mathbf{I}_G$ to the receiver.

B. Decoding Procedure

On the receiver side, after receiving $\mathbf{I}_S$ and $\mathbf{I}_G$, the procedure below is employed using the same shape codebook and gain codebook to reconstruct an output image.

Step 1: For the i-th $(1 \le i \le T)$ shape index, I_{S_i} of $\mathbf{I}_S$, execute the VQ table-lookup procedure (Sect. 6.2.3) to obtain the corresponding shape codeword from $\mathbf{C}_S$. Here let $\mathbf{s}_i$ denote the shape codeword obtained.

Step 2: For the i-th gain index I_{G_i} of $\mathbf{I}_G$, execute the table-lookup procedure again to obtain the corresponding gain value from $\mathbf{C}_G$. Here let g_i be the gain value obtained.

Step 3: Generate an output vector $\mathbf{x}'_i$ from g_i and $\mathbf{s}_i$ using:

$$\mathbf{x}'_i = g_i \times \mathbf{s}_i . \tag{11.1}$$

Step 4: Repeat Step 1 to Step 3 until all the indices have been processed.

Step 5: Assemble all the output vectors $\{\mathbf{x}'_1, \mathbf{x}'_2, ..., \mathbf{x}'_T\}$ to form a reconstructed image $\mathbf{X}'$.

An example of applying the GSVQ procedure to the image of LENA is given in Fig. 11.3, where the sizes of $\mathbf{C}_S$ and $\mathbf{C}_G$ are both 16.

11.2 Watermarking Method

In this section, a watermarking scheme based on GSVQ and visual cryptography is illustrated. A modified visual cryptography is introduced to split the watermark into two meaningless watermarks without expanding their size. The watermarking scheme then hides these two watermarks in the gain indices and shape indices of the GSVQ system and generates two user keys, which can be used later for ownership verification.

11.2.1 Modified Splitting Procedure and Stacking Procedure

In the splitting procedure of the traditional visual cryptography, if the size of a given input image is $M \times N$ pixels, then the sizes of the generated share will be $2M \times 2N$ pixels, as shown in Fig. 11.1. To avoid embedding redundant information on the shares into the cover image, a modified scheme is introduced.

Let $\mathbf{W}$ be a binary image and Table 11.2 be a reference table. The steps below are used to split $\mathbf{W}$ into two meaningless shares.

Table 11.2. The pixel pairs for the modified (2,2)-VSS scheme

Original pixel	□		■	
Number	1	2	3	4
Share 1	□	■	□	■
Share 2	□	■	■	□
Stacked result	□	□	■	■

(a) The original image

(b) The GSVQ coded result

Fig. 11.3. The example of applying GSVQ to the image of LENA (PSNR=28.58 dB)

Step 1: Select a pixel pair randomly from Table 11.2 for each pixel of $\mathbf{W}$. That is, if the given pixel is white, the pair of number 1 or 2 in Table 11.2 is selected randomly. If it is black, then pair number 3 or 4 is selected randomly.

Step 2: Assign the first pixel and the second pixel in the pair obtained to $\mathbf{S}_1$ and $\mathbf{S}_2$ respectively.

Step 3: Repeat these two steps until all the pixels in $\mathbf{W}$ have been processed.

Step 4: Assemble the pixels collected in $\mathbf{S}_1$ to form the first share and the pixels collected in $\mathbf{S}_2$ to form the second share.

In the stacking procedure, the XOR operator is applied to recover the original image:

$$\mathbf{W} = \mathbf{W}_1 \oplus \mathbf{W}_2 . \qquad (11.2)$$

An example of employing the illustrated scheme upon a given image is demonstrated in Fig. 11.4.

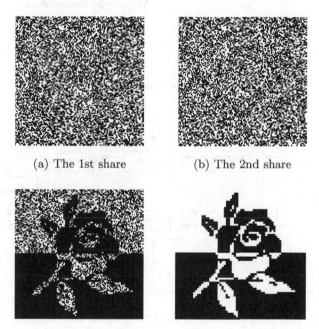

(a) The 1st share   (b) The 2nd share

(c) The stacked result (OR)   (d) The stacked result (XOR)

Fig. 11.4. An example of the modified visual cryptography

11.2.2 Watermarking Algorithm

Based on the modified visual cryptography (Sect. 11.2.1) and the VQ-based watermarking scheme [39] (Sect. 10.1.2), a GSVQ-based watermarking system is illustrated in this section. The block diagrams for illustrating this system are given in Fig. 11.5.

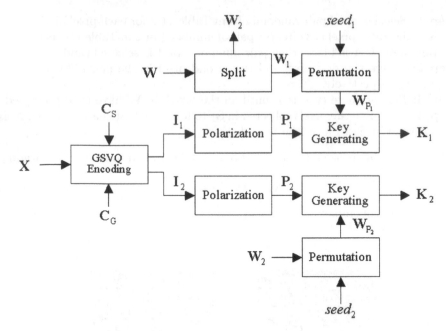

(a) The embedding procedure

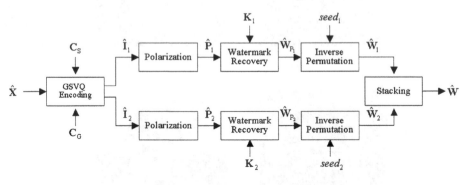

(b) The extraction procedure

Fig. 11.5. The block diagrams of the GSVQ-based watermarking scheme

A. Watermark Embedding Procedure

The steps for the embedding procedure are summarised as:

Step 1: Split the given watermark $\mathbf{W}$ into two shares $\mathbf{W}_1$ and $\mathbf{W}_2$ by applying the method introduced in Sect. 11.2.1.

Step 2: Perform the GSVQ procedure to the cover image $\mathbf{X}$ to obtain the shape index set $\mathbf{I}_1$ and the gain index set $\mathbf{I}_2$.

Step 3: According to $\mathbf{I}_1$ and $\mathbf{I}_2$, generate two polarity streams $\mathbf{P}_1$ and $\mathbf{P}_2$ using the polarization procedure presented in Sect. 10.1.2.

Step 4: Encode $\mathbf{W}_1$ with $\mathbf{P}_1$ and $\mathbf{W}_2$ with $\mathbf{P}_2$ to generate two output keys $\mathbf{K}_1$ and $\mathbf{K}_2$:

$$\mathbf{K}_i = \mathbf{W}_i \oplus \mathbf{P}_i \,, i = 1, 2. \tag{11.3}$$

The generated keys are used for extracting the embedded information and verifying the ownership.

B. Watermark Recovery Procedure

It is not difficult to reveal the original watermark from a watermarked image. We have:

Step 1: Perform the GSVQ procedure on the input image $\hat{\mathbf{X}}$ in order to obtain the shape index set $\hat{\mathbf{I}}_1$ and the gain index set $\hat{\mathbf{I}}_2$.

Step 2: Generate two polarity streams $\hat{\mathbf{P}}_1$ and $\hat{\mathbf{P}}_2$ from $\hat{\mathbf{I}}_1$ and $\hat{\mathbf{I}}_2$, as Step 3 of the embedding procedure.

Step 3: Recover the shadow watermarks $\hat{\mathbf{W}}_1$ and $\hat{\mathbf{W}}_2$ by applying the XOR operator to $\hat{\mathbf{P}}_1$ with $\hat{\mathbf{I}}_1$ and $\hat{\mathbf{P}}_2$ with $\hat{\mathbf{I}}_2$, respectively:

$$\hat{\mathbf{W}}_i = \mathbf{K}_i \oplus \hat{\mathbf{P}}_i \,, i = 1, 2 \,. \tag{11.4}$$

Step 4: Combine $\hat{\mathbf{W}}_1$ and $\hat{\mathbf{W}}_2$ using the method presented in Sect. 11.2.1 to recover the original watermark $\hat{\mathbf{W}}$. If $\hat{\mathbf{X}} = \mathbf{X}'$, which means no artificial modification or natural noise has occurs to $\mathbf{X}'$, then $\hat{\mathbf{W}} = \mathbf{W}$. Otherwise, the recovered watermark $\hat{\mathbf{W}}$ will contain distortion.

11.3   Simulation Results

In the simulation, the image of LENA (Fig. 11.3(a)) was used as the cover image and the image of ROSE (Fig. 11.1(a)) was used as the original watermark respectively. The sizes of them are 512×512 pixels in gray-level and 128×128 pixels in bi-level respectively. The size of the gain codebook and the shape codebook are both 16. The cover image was decomposed into $T = 16384$ blocks of size $d = 4 \times 4$ pixels, and the GSVQ was then applied to these blocks so that the shape indices and the gain indices could be obtained. The watermark was split into two unrecognizable shares, as shown in Figs. 11.4(a) and (b) respectively. They were then hidden in the shape indices and the gain indices respectively. After the embedding procedure, the PSNR (Eq. (2.4)) value between the original image and the GSVQ coded image (Fig. 11.3(b)) is 28.58 dB.

The robustness of the system was tested using some common image-processing procedures, including the JPEG compression with different quality factors (QF), low-pass filtering, median filtering, and rotation. The normalized cross-correlation (NC, see Eq. (2.3)) was used as the evaluating function. Table 11.3 lists the NC values between the input watermarks and the extracted watermarks, and the NC values between the original watermark and the recovered watermarks, respectively. Figure 11.6 demonstrates the final results of the recovered watermarks.

Table 11.3. Robustness test results

Attacks	$NC(\mathbf{W}_1, \hat{\mathbf{W}}_1)$	$NC(\mathbf{W}_2, \hat{\mathbf{W}}_2)$	$NC(\mathbf{W}, \hat{\mathbf{W}})$
JPEG (QF=60%)	0.9765	0.9997	0.9762
JPEG (QF=80%)	0.9893	0.9999	0.9891
Low-pass filtering	0.9648	0.9892	0.9552
Median filtering	0.9614	0.9958	0.9578
Rotation (1°)	0.9355	0.9387	0.8832
Rotation (2°)	0.9309	0.9200	0.8626

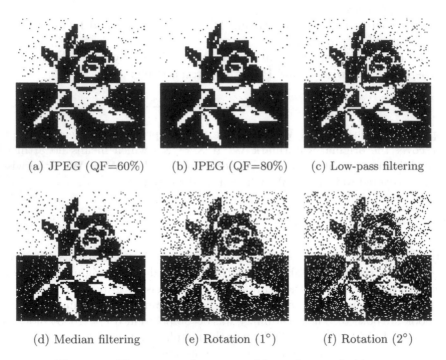

(a) JPEG (QF=60%)   (b) JPEG (QF=80%)   (c) Low-pass filtering

(d) Median filtering   (e) Rotation (1°)   (f) Rotation (2°)

Fig. 11.6. The watermarks recovered from the attacked images

11.4 Summary

In this chapter, a watermarking scheme based on gain-shape VQ and visual cryptography has been introduced. It provides good robustness and imperceptibility, and also improves the shortage of visual cryptography by reducing the size of the encoded images. Experimental results have demonstrated the effectiveness of this watermarking scheme.

Part IV
Summary

Chapter 12
Conclusions and Future Directions

In this chapter, we conclude the book by emphasizing the contribution of the introduced watermarking techniques, and propose some possible further research directions for digital watermarking.

12.1 Conclusions of This Book

In this book, we have introduced and illustrated some important research topics relating to digital watermarking. In the first part, we addressed the motivation of this book in Chap. 1 and introduced the basic fundations of watermarking such as the developement, importances, evaluation, classification, and so on in Chap. 2. Some intelligient techniques, including neural network (NN), evolutionary artificial neural network (EANN), genetic algorithm (GA), and tabu search (TS), were addressed and introduced in Chap. 3. With the background knowledge of watermarking techniques and those intelligient techniques presented, we were able to proceed further to review some watermarking systems introduced in Part 2 and to know how the intelligient techniques could improve the performance.

In Part 2, as we were focusing on the area of performance improvement, several watermarking systems with training techniques were explained. We first introduced the spatial domain based watermarking schemes and a genetic pixel selection (GPS) procedure in Chap. 4. This GPS procedure was applied to find a better set of pixels for embedding. In Chap. 5, a transform domain based watermarking system and a band selection procedure named GBS were presented. By employing the GBS procedure to select the suitable DCT bands for watermaking, the mentioned watermarking system could then provide better performance in visual quality and robustness. In Chap. 6, we introduced several vector quantisation (VQ) based watermarking schemes and a genetic codebook partition (GCP) procedure. This GCP procedure was employed to split the codebook used into two sub-codebooks, so that the embedding results could be improved. In Chap. 7, a genetic index assignment (GIA) procedure was explained. This procedure compressed the watermark and adjusted its signal to suit the cover image, so that a general watermarking scheme could provide larger embedding

F.-H. Wang, J.-S. Pan, and L.C. Jain: Innovations in Dig. Watermark. Tech., SCI 232, pp. 175–178.
springerlink.com

capacity. In Chap. 8, a watermark modification procedure called GWM was introduced. It adjusted the signal of the original watermark and therefore could be employed in general watermarking systems to provide better results in imperceptibility and capacity. From the simulation results of those watermarking schemes and training procedures illustrated in the above chapters, we knew how the watermarking schemes worked and how the training procedure improved the performance.

In Part 3, the focus was shifting to introducing watermarking into some existing image coding systems. We introduced several hybrid systems and pointed out what new abilities the origianl systems could provide. We first presented the multiple description VQ (MDVQ) system and the watermarking scheme based on it in Chap. 9. With the aid of MDVQ, the resistance of the VQ-based watermarking method to channel noise was improved. From another angle of view, with the introduction of the VQ-based watermarking method, a MDVQ system could have the extra ability of watermarking. In Chap. 10, applying the watermarking technique to a multi-stage VQ (MSVQ) system was illustrated. Also, the concept of using fake watermarks was considered. The system not only provided better security for those secrets (the watermarks) embedded, but also added the original MSVQ system the extra value of watermarking. And finally, a watermarking system based on gain-shape VQ (GSVQ) and visual cryptography (VC) was presented in Chap. 11. By the use of VC, the security of the system was improved. From the hybrid systems presented, we learned that new applications or features could be provided by combining several existing systems together. In addition, although the three hybrid systems are all based on VQ, the concept of combining existing coding systems with other domain based watermarking techniques still works.

In summary, we have introduced and illustrated a number of watermarking schemes, training procedures, and hybrid systems in this book. Experimental results and the comparisons to other similar watermarking schemes proposed in literature have shown their effectiveness and usefulness. With the introduction of these systems, we hope this book could help readers in building up the basic ideas and knowledge about digital image watermarking. Further more, with the aid of this book, we hope readers could gain the ability to implement the watermarking systems and the training procedures introduced.

12.2   Future Directions

Watermarking techniques have been developed rapidly in recent years. The term "watermarking" has been seen and heard frequently in many digital-technology-related reports. However, for those people who have been looking forward to using these techniques for solving their problems, it is unfortunate that in the market there is only little software providing such functions. Here the development of artificial intelligent (AI) is borrowed and addressed to explain the reason for the situation mentioned: At the early years of the AI development, researchers and many people optimistically believed that AI could be developed and applied to solve most existing problems. However, with the failures and problems encountered during the development, researchers changed this non-practical idea and focused on designing certain AI techniques to solve certain problems.

For watermarking techniques, similar to AI, there is no perfect algorithm which can be applied to solve all the problems such as copyright protection and ownership verification concurrently. Also, there is no such watermarking algorithm which has the strong resistance against all the possible artificial attacks and natural noise. Therefore, it is not practical to sell products which users will use for different purposes. Currently, researchers are more concerned with designing a particular scheme to solve some certain problems.

In the following section, some points are listed. They can be considered as the possible research directions.

12.2.1 New Watermarking Algorithms

In this category, designing new watermarking algorithms is the focus.

(i) Designing domain-based watermarking algorithms, perhaps using another transform to design a watermarking algorithm is a challenging example.
(ii) Employing cryptography theory to design new asymmetric watermarking schemes or proposing new reversible watermarking schemes are other possible areas. Currently these types of watermarking techniques have not been widely investigated.

12.2.2 New Applications and Utilities

This category includes the research using existing watermarking schemes to solve some certain problems.

(i) As stated in Sect. 2.2, there are many different types of watermarking techniques. How to utilize them to solve some certain problems in real world can be explored.
(ii) Combining some existing systems together to solve certain problems or to provide new useful abilities can be investigated.

12.2.3 Performance Improvement

This category shows the possible directions for improving the performance of existing schemes.

(i) Employing other intelligent techniques to optimize the watermarking schemes can be considered and investigated. It is very likely that these techniques employed may provide better performance than that of those presented in this book or in literature.
(ii) Improving the existing training procedures to shorten the training time or to reduce the system complexity can be studied and investigated.

(iii) Another direction to improve performance is by the design of hybrid systems. It is believed that by combining some existing systems, the shortfalls of one certain system can be improved. Some examples of this concept have been presented in Part 3 of this book. Here we give another two examples: The authors of [105] employed visual cryptography to increase the security of a watermarking system. The authors of [100] proposed some hierarchical-key-design methods to offer the watermarking schemes the ability of sharing secrets with multi-users. Currently, some workshops and international conferences[1] have focused on this area.

12.2.4 Others

This category includes some other non-classified directions.

(i) The study of watermark attacking. For example, how to remove the watermark from the watermarked data without degrading the visual quality of the watermarked data significantly.
(ii) The establishment of standards for watermarking techniques.

---

[1] The International Journal/Conference on Hybrid Intelligent Systems (HIS) is one of journal/conference focuses on hybrid systems.

References

[1] Abraham, A., Köppen, M., Franke, K. (eds.): Design and Application of Hybrid Intelligent Systems. IOS Press, Amsterdam (2003)

[2] Ahmed, N., Natarajan, T., Rao, K.R.: Discrete cosine transforms. IEEE Trans. on Comp. C-23, 90–93 (1974)

[3] Barni, M., Bartolini, F.: Watermarking Systems Engineering: Enabling Digital Assets Security and Other Applications. Marcel Dekker, Inc., New York (2004)

[4] Cheung, W.N.: Digital image watermarking in spatial and transform domains. In: Proc. IEEE TENCON 2000, vol. 3, pp. 374–378 (2000)

[5] Cole, E.: Hiding in Plain Sight: Steganography and the Art of Covert Communication. Wiley Publishing, Inc., Indiana (2003)

[6] Charrier, M., Cruz, D.S., Larsson, M.: JPEG2000, the next millennium compression standard for still images. In: IEEE Int. Conf. on Multimedia Computing and Systems, vol. 1, June 1999, pp. 131–132 (1999)

[7] Carpenter, G.A., Grossberg, S.: A massively parallel architecture for a self-organizing neural pattern recognition machine. Computer Vision, Graphics and Image Processing 37, 54–115 (1987)

[8] Carpenter, G.A., Grossberg, S., Rosen, D.B.: Fuzzy ART: Fast stable learning and categorization of analog ptterns by an adaptive resonance system. Neural Networks 4, 759–771 (1991)

[9] Chu, S.C., Hsin, Y.C., Huang, H.C., Huang, K.C., Pan, J.S.: Multiple description watermarking for lossy network. In: IEEE Int. Symposium on Circuits and Systems, May 2005, vol. 4, pp. 3990–3993 (2005)

[10] Cox, I.J., Kilian, J., Leighton, F.T., Shamoon, T.: Secure spread spectrum watermarking for multimedia. IEEE Trans. on Image Proc. 6(12), 1673–1687 (1997)

[11] Cox, I.J., Miller, M.I., Bloom, J.A.: Digital Watermarking, 2nd edn. Morgan Kaufman Publishers, San Francisco (2000)

[12] Christopoulos, C., Skodras, A., Ebrahimi, T.: The JPEG2000 still image coding system: an overview. IEEE Trans. on Consumer Electronics 46(4), 1103–1127 (2000)

[13] Chung, Y.Y., Wong, M.T.: Implementation of digital watermarking system. In: IEEE Int. Conf. on Consumer Electronics (ICCE 2003), June 2003, pp. 214–215 (2003)

[14] Chang, C.H., Ye, Z., Zhang, M.: Fuzzy-ART based adaptive digital watermarking scheme. IEEE Trans. on Circuits and Systems for Video Technology 15, 65–81 (2005)

[15] Du, W.-C., Hsu, W.-J.: Adaptive data hiding based on VQ compressed images. IEE Proc.-Vis. Image Signal Process. 150(4), 233–238 (1999)

[16] Dumitrescu, D., Lazzerini, B., Jain, L.C., Dumitrescu, A.: Evolutionary Computing and Applications. CRC Press, USA (2000)

[17] Engelbrecht, A.P.: Computational Intelligence: An Introduction. John Wiley & Sons Inc., Chichester (2002)

[18] Fu, M.S., Au, O.C.: Joint visual cryptography and watermarking. In: IEEE Int. Conf. on Multimedia and Expo. (ICME 2004), Taipei, June 2004, p. 4 (2004)

[19] Grossberg, S.: Adaptive pattern classification and universal recoding, I: Parallel development and coding of neural feature detectors. Biological Cybernetics 23, 121–134 (1976)

[20] Grossberg, S.: Adaptive pattern classification and universal recoding, II: Feedback, expectation, olfaction, and illusions. Biological Cybernetics 23, 187–202 (1976)

[21] Goldberg, D.E.: Genetic Algorithms in Search, Optimization and Machine Learning. Addison-Wesley, Massachusetts (1992)

[22] Goyal, V.K.: Multiple description coding: Compression meets the network. IEEE Signal Processing Magazine 18(5), 74–93 (2001)

[23] El Gamal, A.A., Cover, T.M.: Achievable rates for multiple descriptions. IEEE Trans. Ingorm Theory 28, 851–857 (1982)

[24] Gersho, A., Gray, R.M.: Vector Quantization and Signal Compression. Kluwer Academic Publisher, London (1992)

[25] Glover, F., Laguna, M.: Tabu Search. Kluwer Academic Publishers, Boston (1997)

[26] Görtz, N., Leelapornchai, P.: Optimization of the index assignments for multiple description vector quantizers. IEEE Trans. Commun. 51(3), 336–340 (2003)

[27] Gonzalez, R.C., Woods, R.E.: Digital Image Processing. Addison-Wesley, MA (1992)

[28] Huang, H.-C.: Genetic-Based Algorithms for Image Compression Classification and Watermarking, PhD Thesis (2001)

[29] Hwang, J.J.: Digital image watermarking employing codebook in vector quantisation. IEE Electronics Letters 39(11), 840–841 (2003)

[30] Hu, Y.C., Chang, C.C.: A progressive codebook training algorithm for image vector quantization. In: Proceedings of Fifth Asia-Pacific Conference on Communications and Fourth Optoelectronics and Communications Conference (1999), October 1999, vol. 2, pp. 936–939 (1999)

[31] Hou, Y.C., Chen, P.M.: An asymmetric watermarking scheme based on visual cryptography. In: IEEE Proc. 5th ICSP, pp. 992–995 (2000)

[32] Hwang, M.S., Chang, C.C., Hwang, K.F.: Digital watermarking of images using neural networks. Journal of Electronic Imaging 9, 548–555 (2000)

[33] Hernández, J.R., Pérez-González, F., Rodriguez, J.M.: The impact of channel coding on the performance of spatial watermarking for copyright protection. In: Proc. IEEE Int. Conf. Acoustics, Speech, and Signal Processing, pp. 2973–2976

[34] Huang, H.-C., Pan, J.S., Wang, F.H.: An embedding algorithm for multiple watermarks. In: Proc. 6th Int. Conf. on Knowledge-Based Intelligent Information and Engineering System (KES 2002), Crema, Italy, pp. 412–416. IOS Press, Amsterdam (2002)

[35] Huang, H.-C., Pan, J.S., Wang, F.H.: Genetic watermarking on transform domain. In: Pan, J.S., et al. (eds.) Intelligent Watermarking Techniques, ch. 12, pp. 351–376. World Scientific, Singapore (2004)

[36] Huang, H.-C., Pan, J.S., Wang, F.H., Shieh, C.S.: Robust image watermarking with tabu search approaches. In: IEEE Int. Symposium on Consumer Electronics (ISCE 2003), Sydney (October 2003)

[37] Hsu, C.T., Wu, J.L.: Hidden digital watermarks in images. IEEE Trans. on Image Proc. 8, 58–68 (1999)

[38] Huang, C.-H., Wu, J.-L.: Attacking visible watermarking schemes. IEEE Trans. on Multimedia 6(1), 16–30 (2004)

[39] Huang, H.-C., Wang, F.H., Pan, J.S.: Efficient and robust watermarking algorithm with vector quantisation. IEE Electronics Letters 37(13), 826–828 (2001)

[40] Huang, H.-C., Wang, F.H., Pan, J.S.: A VQ-based robust multi-watermarking algorithm. IEICE Trans. on Fundamentals of Electronics, Communication and Computer Sciences E85-A(7), 1719–1726 (2002)

[41] Jain, L.C. (ed.): Soft Computing Techniques in Knowledge-Based Intelligent Engineering Systems. Springer, Germany (1997)

[42] Jain, L.C.: Integration of Neural Net, Fuzzy Systems and Evolutionary Computing in System Design. In: Proceedings of the Third Asian Fuzzy Systems Symposium, Masan, Korea, pp. 28–30 (1998)

[43] Jo, M., Kim, H.: A digital image watermarking scheme based on vector quantisation. IEICE Trans. on Inf. & Syst. E85-D, 1054–1056 (2002)

[44] Je, S.-K., Seo, Y.-S., Lee, S.-J., Cha, E.-Y.: Self-organizing coefficient for semi-blind watermarking. In: Zhou, X., Zhang, Y., Orlowska, M.E. (eds.) APWeb 2003. LNCS, vol. 2642, pp. 275–286. Springer, Heidelberg (2003)

[45] Kohonen, T.: Self-organizing formation of topologically correct feature maps. Biological Cybernetics 43, 59–69 (1982)

[46] Kohonen, T.: Self-Organization and Associative Memory. Springer, Berlin (1984)

[47] Katzenbeisser, S., Petitcolas, F. (eds.): Information Hiding — Techniques for Steganography and Digital Watermarking. Artech House, Norwood (2000)

[48] Kohonen, T., Somervuo, P.: How to make large self-organizing maps for nonvectorial data. Neural Networks 15, 945–952 (2002)

[49] Laguna, M.: A guide to implementing tabu search. Investigación Operativa 4(1), 5–25 (1994)

[50] Loo, P., Kingsbury, N.: Watermark detection based on the properties of error control codes. IEE Proceedings- Vision, Image and Signal Processing 150(2), 115–121 (2003)

[51] Lu, Z.M., Pan, J.S., Sun, S.H.: VQ-based digital image watermarking method. IEE Electronics Letters 36(14), 1201–1202 (2000)

[52] Lu, Z.M., Sun, S.H.: Digital image watermarking technique based on vector quantisation. IEE Electronics Letters 36(4), 303–305 (2000)

[53] Langelaar, G.C., Setyawan, I., Lagendijk, R.L.: Watermarking digital image and video data: A state-of-the-art overview. IEEE Signal Processing Magazine 17, 20–46 (2000)

[54] Lin, C.-Y., Wu, M., Bloom, J.A., Cox, I.J., Miller, M.L., Lui, Y.M.: Rotation, scale, and translation resilient watermarking for images. IEEE Trans. on Image Processing 10(5), 767–782 (2001)

[55] Moore, B.: ART 1 and Pattern Clustering. In: Touretzky, D., Hinton, G., Sejnowski, T. (eds.) Proceedings of the 1988 Connectionist Models Summer School, San Mateo, CA, pp. 174–185. Morgan Kaufmann, San Francisco (1989)

[56] Malvar, H.S.: Signal Processing with Lapped Orthogonal Transforms. Artech House, Norwood (1992)

[57] Mitchell, O.R., Delp, E.J., Carlton, S.G.: Block truncation coding: a new approach to image compression. In: Proc. ICC (1978)

[58] Mitrokotsa, A., Komninos, N., Douligeris, C.: Intrusion detection with neural networks and watermarking techniques for MANET. In: IEEE Int. Conf. on Pervasive Services, July 2007, pp. 118–127 (2007)

[59] Nikolaidis, N., Pitas, I.: Digital image watermarking: an overview. In: IEEE Int. Conf. on Multimedia Computing and Systems, vol. 1, pp. 1–6 (1999)

[60] Noar, M., Shamir, A.: 'Visual cryptography. LNCS, pp. 1–12. Springer, Heidelberg (1994)

[61] Proakis, J.G.: Digital Communications, 3rd edn. McGraw-Hill, New York (1995)

[62] Petitcolas, F.A.P.: Watermarking schemes evaluation. IEEE Signal Processing Magazine 17(5), 58–64 (2000)

[63] Petitcolas, F.A.P.: Stirmark benchmark 4.0, (February 2009) http://www.petitcolas.net/fabien/watermarking/stirmark/

[64] CERTIMARK, Certification for Watermarking Techniques (EU Project IST-1999-10987) (February 2009), http://www.certimark.org

[65] Pan, J.S., Hsin, Y.C., Huang, H.C., Huang, K.C.: Robust image watermarking based on multiple description vector quantization. Electronics Letters 40(22), 1409–1410 (2004)

[66] Pan, J.S., Huang, C.Y., Huang, H.C., Liao, B.Y., Huang, K.C.: Tabu search based multi-watermarks embedding algorithm with multiple description coding. Elsevier Science, Amsterdam (2005)

[67] Pan, J.S., Huang, H.-C., Jain, L.C. (eds.): Intelligent Watermarking Techniques. World Scientific, Singapore (2004)

[68] Pan, J.S., Huang, H.-C., Wang, F.H.: Genetic watermarking techniques. In: Proc. 5th Int. Conf. on Knowledge-Based Intelligent Information Engineering Systems & Allied Technologies (KES 2001), Osaka, Japan, September 2001, pp. 1032–1036 (2001)

[69] Pereira, C., Ruanaidh, J.J.K.Ó., Pun, T.: Secure robust digital watermarking using the lapped orthogonal transform. In: IS&T/SPIE Electronic Imaging 1999, San Jose, CA, USA (January 1999)

[70] Pereira, S., Voloshynovskiy, S., Madueno, M., Marchand-Maillet, S., Pun, T.: Second generation benchmarking and application oriented evaluation. In: Information Hiding Workshop III, pp. 340–353 (2001)

[71] Pan, J.S., Wang, F.H., Huang, H.-C., Jain, L.C.: Improved schemes for VQ-based image watermarking. In: IEEE Int. Symposium on Consumer Electronics (ISCE 2003), Sydney (December 2003)

[72] Jeng-Shyang, P., Feng-Hsing, W., Lakhmi, J., Nikhil, I.: A multistage VQ based watermarking technique with fake watermarks. In: Petitcolas, F.A.P., Kim, H.-J. (eds.) IWDW 2002. LNCS, vol. 2613, pp. 81–90. Springer, Heidelberg (2003)

[73] Pan, J.S., Wang, F.H., Yang, T.C., Jain, L.C.: A gain-shape VQ based watermarking technique with modified visual secret sharing scheme. In: Proc. 6th Int. Conf. on Knowledge-Based Intelligent Information and Engineering Systems (KES 2002), Crema, Italy, September 2002, pp. 402–406. IOS Press, Amsterdam (2002)

[74] Podilchuk, C.I., Zeng, W.J.: Image-adaptive watermarking using visual models. IEEE Journal on Selected Areas in Communications 16(4), 525–539 (1998)

[75] Stinson, D.: Visual cryptography and threshold schemes. IEEE Potentials 18, 13–16 (1999)

[76] Shieh, C.S., Huang, H.-C., Wang, F.H., Pan, J.S.: An embedding algorithm for multiple watermarks. Journal of Information Science and Engineering (JISE) 19(2), 381–395 (2003)

[77] Shieh, C.S., Huang, H.-C., Wang, F.H., Pan, J.S.: Genetic watermarking based on transform-domain techniques. Pattern Recognition 37(3), 555–565 (2004)

[78] Suhail, M.A., Obaidat, M.S.: On the digital watermarking in JPEG 2000. In: Proc. 8th IEEE Int. Conf. on Electronics, Circuits and Systems (ICECS 2001), September 2001, vol. 2, pp. 871–874 (2001)

[79] Suhail, M.A., Obaidat, M.S.: Digital watermarking-based DCT and JPEG model. IEEE Trans. on Instrumentation and Measurement 52(5), 1640–1647 (2003)

[80] Solachidis, V., Tefas, A., Nikolaidis, N., Tsekeridou, S., Nikolaidis, A., Pitas, I.: A benchmarking protocal for watermarking methods. In: Proc. Int. Conf. Image Processing, pp. 1023–1026 (2001)

[81] van Schyndel, R.G., Tirkel, A.Z., Osborne, C.F.: A digital watermark. In: Proc. IEEE Int. Conf. Image Processing (ICIP 1994), November 1994, vol. 2, pp. 86–90 (1994)

[82] Tirkel, A.Z., Rankin, G.A., van Schyndel, R.M., Ho, W.J., Mee, N.R.A., Osborne, C.F.: Electronic water mark. In: Digital Image Computing Techniques and Applications 1993, pp. 666–672 (1993)

[83] US Patent 6718045 - Method and Device for Inserting a Watermarking Signal in an Image

[84] Vaishampayan, V.A.: Design of multiple description scalar quantizers. IEEE Trans. Inf. Theory 39(3), 821–834 (1993)

[85] De Vleeschouwer, C., Delaigle, J.-F., Macq, B.: Invisibility and application functionalities in perceptual watermarking—An overview. Proc. of the IEEE 90(1), 64–77 (2002)

[86] Vonk, E., Jain, L.C., Veelenturf, L.P.J., Johnson, R.: Generation of a Neural Network Architecture using Evolutionary Computation. In: ETD 2000, pp. 142–147. IEEE Computer Society Press, USA (1995)

[87] Vetterli, M., Kovačević, J.: Wavelets and Subband Coding. Prentice-Hall Inc., Englewood Cliffs (1995)

[88] Voyatzis, G., Pitas, I.: Chaotic watermarks for embedding in the spatial digital image domain. In: Proc. IEEE Int'l Conf. Image Processing, pp. 432–436 (1998)

[89] Voloshynovskiy, S., Pereira, S., Pun, T., Eggers, J.J., Su, J.K.: Attacks on digital watermarks: classification, estimation based attacks, and benchmarks. IEEE Communications Magazine 39(8), 118–126 (2001)

[90] Wang, H.J., Chang, C.Y., Pan, S.W.: A DWT-based robust watermarking scheme with fuzzy ART. In: Int. Joint Conf. on Neural Networks (IJCNN 2006), pp. 1750–1757 (2006)

[91] Wolfgang, R.B., Delp, E.J.: Overview of image security techniques with applications in multimedia systems. In: Proc. SPIE Conf. Multimedia Networks: Security, Displays, Terminals, and Gateways, pp. 297–308 (1997)

[92] Wang, F.H., Jain, L.C., Pan, J.S.: Genetic watermarking techniques based on spatial domain. In: Proc. 6th Int. Conf. on Knowledge-Based Intelligent Information and Engineering System (KES 2002), Crema, Italy, September 2002, pp. 417–422. IOS Press, Amsterdam (2002)

[93] Wang, F.H., Jain, L.C., Pan, J.S.: A novel VQ-based watermarking scheme with genetic codebook partition. In: Proc. 3rd Int. Conf. on Hybrid Intelligent Systems (HIS 2003), Melbourne, December 2003, pp. 1003–1011. IOS Press, Amsterdam (2003)

[94] Wang, F.H., Jain, L.C., Pan, J.S.: Genetic watermarking on spatial domain. In: Pan, J.S., et al. (eds.) Intelligent Watermarking Techniques, March 2004, ch. 13, pp. 377–393. World Scientific, Singapore (2004)

[95] Wang, F.H., Jain, L.C., Pan, J.S.: Watermark embedding system based on visual cryptography. In: Pan, J.S., et al. (eds.) Intelligent Watermarking Techniques, March 2004, ch. 17, pp. 481–513. World Scientific, Singapore (2004)

[96] Wang, F.H., Jain, L.C., Pan, J.S.: VQ-based gray watermark embedding scheme with genetic index assignment. Int. Journal of Computational Intelligence and Applications (IJCIA) 4(2), 165–181 (2004)

[97] Wang, F.H., Jain, L.C., Pan, J.S.: A novel VQ-based watermarking scheme with genetic codebook partition. Journal of Network and Computation Applications (JNCA) 30(1), 4–23 (2007)

[98] Wang, F.H., Jain, L.C., Pan, J.S.: Genetic watermark modification for watermarking schemes. In: Proc. of 2nd Int. Conf. on Artificial Intelligence in Science and Technology (AISAT 2004), Hobart, Australia, November 2004, pp. 196–199 (2004)

[99] Wang, F.H., Jain, L.C., Pan, J.S.: Hiding watermark in watermark. In: 2005 IEEE Int. Symposium on Circuits and Systems Technology (ISCAS 2005), Kobe, Japan, May 2005, pp. 4975–V4978 (2005)

[100] Wang, F.H., Pan, J.S., Jain, L.C.: Shadow watermark embedding system. In: 2005 IEEE Int. Symposium on Circuits and Systems Technology (ISCAS 2005), Kobe, Japan, May 2005, pp. 4018–V4021 (2005)

[101] Wang, F.H., Pan, J.S., Jain, L.C., Huang, H.-C.: A VQ-based image-in-image data hiding scheme. In: IEEE Int. Conf. on Multimedia and Expo. (ICME 2004), Taipei, June 2004, p. 4 (2004)

[102] Wang, F.-H., Pan, J.-S., Jain, L., Huang, H.-C.: VQ-based gray watermark hiding scheme and genetic index assignment. In: Aizawa, K., Nakamura, Y., Satoh, S. (eds.) PCM 2004. LNCS, vol. 3332, pp. 73–80. Springer, Heidelberg (2004)

[103] Wang, Y., Orchard, M.T., Vaishampayan, V.A., Reibman, A.R.: Multiple description coding using pairwise correlating transforms. IEEE Trans. Image Processing 10(3), 351–366 (2001)

[104] Wang, Y., Reibman, A.R., Orchard, M.T., Jafarkhani, H.: An improvement to multiple description transform coding. IEEE Trans. Signal Processing 50(11), 2843–2854 (2002)

[105] Wang, C.C., Tai, S.C., Yu, C.S.: Repeating image watermarking technique by the visual cryptography. In: IEICE Special Section on Digital Signal Processing, pp. 1589–1598 (2000)

[106] Yao, X.: Evolving artificial neural networks. Proceedings of IEEE 87, 1423–1447 (1999)

[107] Yao, X.: Evolutionary computation: theory and applications. World Scientific, Singapore (1999)

[108] Zhang, Z., Berger, T.: New results in binary multiple descriptions. IEEE Trans. Inform. Theory 33(4), 502–521 (1987)

Index